Philosophy of
science and sociology

International Library of Sociology

Founded by Karl Mannheim

Editor: John Rex, University of Aston in Birmingham

Arbor Scientiae
Arbor Vitae

A catalogue of the books available in the **International Library of Sociology** and other series of Social Science books published by Routledge & Kegan Paul will be found at the end of this volume.

Philosophy of science and sociology

From the methodological doctrine to research practice

Edmund Mokrzycki

Routledge & Kegan Paul
London, Boston, Melbourne and Henley

First published in English in 1983
by Routledge & Kegan Paul plc
39 Store Street, London WC1E 7DD,
9 Park Street, Boston, Mass. 02108, USA,
296 Beaconsfield Parade, Middle Park,
Melbourne, 3206, Australia and
Broadway House, Newtown Road,
Henley-on-Thames, Oxon RG9 1EN
Set by Columns, Reading
Printed in Great Britain by
St. Edmundsbury Press, Suffolk

Library of Congress Cataloging in Publication Data

Mokrzycki, Edmund.

Philosophy of science and sociology.
International library of sociology
Translation of: Filozofia nauki a socjologia.
Includes bibliographical references and index.
1. Science - Philosophy. 2. Science - Methodology.
3. Science - Social aspects. I. Title. II Series.
Q175.M7313 1983   501   82-18624

ISBN 0-7100-9444-2

Contents

1 Introduction

> Logicians cannot make sense of science - but they can make
> sense of logic and so they stipulate that science must be
> presented in terms of their favourite logical system. This
> would be excellent comedy material were it not the case that
> by now almost everyone has started taking the logicians
> seriously.(1)

It is difficult to guess to which population that almost everyone
was intended by Feyerabend to apply, but it would be easy to
show that the seriousness with which the various academic
milieus treat the logic of science is a gradeable feature and that
there are in that respect essential differences between, for
instance, the milieu of physicists and that of sociologists. In
the present book I am concerned, among other things, with the
attitude of sociologists qua a certain group toward the logic
of science, and especially with the objectified form of that attitude, that is with the scholarly work of sociology.

Instead of the term logic of science I shall use the seemingly
broader term philosophy of science. Of course, the latter term
can, and sometimes is, interpreted broadly as all kinds of philosophical reflection on science, but usually it refers to only one
way of philosophizing about science, namely that which originated with the Vienna Circle and other centres having a similar
orientation (the Berlin School, a group of British analytic
philosophers and several other small groups in various countries,
primarily in the United States, the Netherlands, and the
Scandinavian countries)(2) and which came to single out logical
positivism as a trend in philosophy. In other words, *the philosophy of science* functions as a historical term which applies to
a specified trend in the history of thought, and will be used in
that sense here. This is additionally legitimated by the fact
that the said trend has developed into an academic discipline:
the philosophy of science as it functions today as an academic
discipline is the product of a single orientation, and its origin
is clearly connected with the trend of 'scientific philosophy' in
the 1920s (the first international congress organized by the
Vienna Circle in Prague, in 1929, can well be regarded as the
first congress of representatives of the new discipline). Of
course, during its fifty years' history, the philosophy of science
has evolved considerably. There have also always been differences of opinion among its representatives. This, however,
does not change the fact that among the possible, and even

actually existing, types of philosophical reflection on science, the philosophy of science has represented a coherent whole: the differences of opinion between, for example, Carnap and Popper as compared with the views of, for example, Ingarden or Habermas, referred to secondary issues.

Early logical positivism primarily imposed on the subject matter and the method of research on the philosophy of science. Lip service was paid to covering science as such, but in fact research came to be confined to that sphere of phenomena in the modern history of Western civilization which academic circles in the 1920s labelled as science. The research method was defined as the logical analysis of the language of science and later, of products of science. This in fact proved to be an analysis of products of science (or what were considered to be products of science) in the light of contemporaneous logic. This is why the terms the logic of science and the philosophy of science can be treated as interchangeable. We shall revert to these issues on many occasions. For the time being let me just mention the fact that the philosophy of science as we have it today was based on specified assumptions and that the adoption of those assumptions came to determine its later evolution, including the latest studies which seemingly mark a turning point in that discipline.

Thus, when I refer to the philosophy of science, I mean that very coherent whole, clad in the paraphernalia of an academic discipline, distinguishable in philosophical reflection on science, which is closely linked to logical positivism. It does not cover philosophical reflection on science as practised by such authors as Karl Mannheim, Alfred Schutz, Michael Polanyi, Peter Winch and Jürgen Habermas. Thus I do not mean all philosophical reflection on science, but only its dominant (in the institutional sense of the word) form, organized around a certain philosophical doctrine.

The subject matter with which I am concerned here covers the impact of the philosophy of science understood in that way, on sociology and related disciplines, and the main thesis of the present book can be formulated thus: the impact of the philosophy of science on sociology over the last fifty years, i.e., from the rise of the Vienna Circle, has resulted in a deep-reaching and basically undesirable methodological reorientation in sociology. The turning point comes in the 1940s and 1950s. At that time, radical and very restrictive methodological ideas that marked the early period of logical positivism, i.e., ideas shaped in the 1920s and 1930s, in sociology won the status of the scientific interpretation of scientific method and gradually came to circulate as the methodological foundation of sociology. That discipline accordingly lost much of its humanistic nature without, however, contrary to current opinions, acquiring the status of a scientific discipline in the narrower sense of the term, because sociology came not so much to be made scientific, but to be shaped after a methodological model of science which had already been abandoned as inadequate in the discipline where it originated, that is, in the philosophy of science.

Contrary to appearances, this claim remains outside 'anti-positivism' and 'antiscientism', at least if these terms are interpreted in accordance with the way they are used in reflection on sociology, although it is connected with these two trends by the subject matter of criticism. That subject matter is, of course, the main trend in present-day sociology, at one time called 'contemporary empirical sociology' or 'empiricist sociology', and today more and more often, especially in the United States, called 'traditional sociology'. Every competent sociologist can easily draw a fairly clear demarcation line between that trend and the rest of sociology, even though defining the place of that trend in the discipline as a whole and the answer to the question, how far sociology has been dominated by that trend, are questions of intuitive estimates and depend, among other things, on contacts and links with the various milieus: Jürgen Habermas, who is comparatively little connected with academic sociology, is less inclined to see the omnipresent domination of empiricist sociology (if we agree to use this conventional term) than was Stanisław Ossowski, at one time President of the Board of the Polish Sociological Association and Vice-President of the International Sociological Association. I do not, it seems, risk much by claiming, in the present book, that while there are no grounds for equating empiricist sociology with present-day academic sociology, the domination of empiricist sociology is strong enough to make its imprint upon the discipline as a whole.

In a sense, this is due to change in the linguistic usage: the concept of sociology has changed, so that part of the work which traditionally used to be included in sociology, at a certain point moved outside the scope of that conceptual category. This happened, for instance, to reflection on social problems, based on the Frankfurt School's approach. Such work has not been abandoned, even though it has left the field associated with the term sociology. It might seem, therefore, that the domination of sociology by the said trend is, in fact, apparent rather than real, a process which took place in the sphere of language rather than in that of research. The point is, however, that those changes in linguistic usage merely reflected what was taking place in the discipline itself; to be more specific, the extension of the term sociology was brought into line with what organized academic activity treated as sociology. Hence, if part of the output which was traditionally treated as sociological is now outside the scope of the term sociology or has an obscure status relative to the scope of that term, this is usually due to the fact that it has been driven outside sociology or removed to the peripheries of that discipline. The struggle for a place within sociology is, of course, a struggle for a place within that discipline, and not for a place in a certain classification of science.

Anyone who has followed, even if only cursorily, the evolution of sociology over the last decades, can easily see that the struggle for a place within that discipline was exceptionally vehement, at times even brutal, and that it rarely complied with the current conception of science as the arena on which ideas clash and the truth emerges victorious. From the point of view of the sociology of science, such a picture is naive regardless of the academic discipline we have in mind.(3) But sociology, together with related disciplines, such as psychology and political science, is in an exceptional position: it is a discipline in which the very status of being scientific is at stake, and not just position in the hierarchy of scientific schools. Defeat in conflict may mean an immediate elimination from the discipline or being assigned the status of an intellectual relic. Victory opens the way to acquiring a monopoly on being scientific within the discipline, which in practice means having the status of 'the only scientific', 'truly scientific', 'the most scientific', or just 'scientific' orientation.

Just such a position was won by the empiricist trend in sociology. This was in a sense natural, for that trend was born of 'the quest to make sociology scientific', and its success, whether real or apparent, automatically assigned to those engaged in that quest the rank of leading scholars in the field. But all the rest were thereby, also automatically assigned the ambiguous status of representatives of pre-scientific sociology. The very rise of scientific sociology assumes the existence of non-scientific sociology. The same applies to empirical sociology: its reason for existence as a separate trend is based on the non-empirical nature of the other trends. Regardless of the underlying intentions and the arguments used, 'contemporary empirical sociology' won its position by questioning the raison d'être of the other trends and theoretical and methodological orientations. Its birth and development was thus necessarily accompanied by impassioned polemics, in which one side was blamed for being non-scientific, and the other, for being pseudo-scientific, thus in effect, for the same thing. Lundberg, with great rhetorical and polemical skill, showed that classical sociology was based on methodological obscurantism; Sorokin did not conceal his conviction that the trend meant 'to make sociology scientific' (represented among others, by Lundberg) was nothing more than academic chutzpah. It is not true that all of the controversy over empiricist sociology was conducted in such a spirit, but the fact remains that the *meaning* of the arguments used in most cases intended placing the opponent outside the limits of 'true' sociology, and thus de facto questioned his scholarly competence.

The controversy has never died down, even though in the mid-1960s it seemed that the matter has been settled in favour of 'empirical sociology'. This was because that trend triumphed on the social level (curricula, research programs, activity of learned societies, approval on the part of university authorities and foundations, etc.), but not on the intellectual one. The arguments advanced by its opponents, which questioned the theoretical foundations and the intellectual values of the whole undertaking, have never been refuted. It suffices to realize that such publications as 'Methods in Sociology' by Ellwood, 'Fads and Foibles in Modern Sociology' by Sorokin, 'The Sociological Imagination' by Mills, and 'O osobliwościach nauk społecznych' ('On the Peculiarities of the Social Sciences') by Ossowski have, in fact, remained unanswered.(4) And yet in those publications, much publicized at one time and quite influential in certain spheres, the assumptions on which empiricist sociology was based were literally annihilated. Hayek's well-known 'pamphlet' aimed at 'scientism' in the social disciplines, i.e., 'The Counter Revolution in Science',(5) also never received a thorough and deep answer. The same applies to the critical opinions voiced by Znaniecki, Blumer, MacIver, Adorno, Schutz, and many other sociologists, who represented various theoretical traditions and methodological orientations. The criticism of 'empirical sociology' and organized academic life each went their own way. Such a volume of unanswered critical arguments has accumulated during the last fifty years that even a small part of it would suffice to make one conclude that establishing 'the only scientific' sociology is an untenable undertaking. Yet it has turned out that even the most profound and convincing arguments are powerless if aimed at something which is popular and in agreement with the spirit of the times.

That spirit of the times was partly modified in the 1970s, but not enough to result in a quick undermining of the dominant position of 'empirical sociology'. Attacks against the sociological 'establishment', the feeling of a crisis in sociological knowledge, the revival of some old, and the birth of some new, theoretical and methodological trends and orientations have changed the atmosphere in sociology, but not greatly affected the balance of power. 'Empirical sociology' is so strongly entrenched in academic life that its dominant position is guaranteed for many years to come by the vested interests of thousands of people. It is true that almost every major American university strives to have one or two ethnomethodologists and to offer courses in 'qualitative' (versus 'quantitative') methodology, but this merely proves that a prestigious academic institution cannot lightly dismiss intellectual novelties, even risky ones.

Yet the 1970s have achieved what Ernest Becker has termed a 'rehabilitation of scientific debate'. It is true that it took place following grave disturbances to academic life in the late 1960s, and early 1970s, but when disorganization in academic teaching came to an end, it became more and more obvious that coming out against 'traditional' sociology could not be equated with intellectual trouble-making, that apart from intellectually futile and irresponsible acts there was vigorous, but relevant and deep-reaching criticism. These two things can today easily be separated from one another, nor is there any doubt that the existence of such trends as ethnomethodology and humanistic sociology is not a transient phenomenon.(6) While a single socio-logical orientation is still strongly dominant in the life of academic institutions, sociological production is marked by a clear and growing methodological and theoretical pluralism. It is that pluralism which, even though it lacks institutional sup-port, has restored the social justification of questions about the methodological and theoretical nature of 'empirical sociology'.

Stanisław Ossowski thought that 'empirical sociology' shaped its postulates after the pattern of natural science, and added that this applied to research methods, the criteria of what is scienti-fic, and to practical applications as well.(7) His well-known paper (later included as a chapter in 'O osobliwościach nauk społecznych'), from which this opinion is drawn, is called, Wzory nauk przyrodniczych w empirycznej socjologii (Patterns of Natural Science in Empirical Sociology). A similar character-istic of 'empirical sociology' can be found in most critics of that trend, beginning with Mills, Schutz, and Znaniecki, and ending in young present-day interactionists, ethnomethodologists, radicals, and humanists. Furthermore, this characteristic is also accepted by protagonists of 'empirical sociology'. Agree-ment about diagnosis is thus almost universal, although the evaluations of that fact, i.e., the application of patterns drawn from natural science, to use Ossowski's terminology, differ.

   In my opinion, that diagnosis is erroneous; it is also dangerous in so far as it makes the discussion of the methodological foundations of sociology take a wrong course. 'Empirical socio-logy' takes its patterns neither from natural science in general, nor physics in particular, nor even, as some critics would have it, classical mechanics, but is an attempt to create a science of society that would be in agreement with certain *ideas* about the methodological characteristics of natural science. The mis-take that most critics of the empiricist trend in sociology make, is to be too quick to admit that it is related to modern natural science in general, and to physics in particular. The relation-ship is purely hypothetical, and without analyzing that hypo-thesis one cannot properly judge what has happened in sociology during the last fifty years. As long as 'empirical sociology' has the authority of physics behind it, discussion of the applic-ability of 'patterns drawn from natural science' will of necessity focus on the peculiarities of the social sciences while disregard-ing the peculiarity of those patterns.

   The present book is an attempt to analyze the issue of 'patterns drawn from natural science', as they function in 'empirical sociology'. Unfortunately I can do this only on a very limited scale. An exhaustive study would require the co-opera-tion of competent representatives of natural science and would result in a publication of a different kind. On the other hand, I am not sure that such a vast undertaking is really necessary in this case. As I will try to show, the empiricist trend in

sociology was methodologically based on a system of opinions
that were a conglomerate of distorted elements of the methodological doctrine of logical positivism and popular opinions on
the nature of scientific knowledge. The properties of that
conglomerate are of a kind that force us to preclude the possibility of physics attaining its present level on such a methodological basis. Obviously, the physicists have the last say on
this issue, but it seems that they pose no threat to the claim
that 'empirical sociology' is pseudo-scientific.

The present book is thus not a criticism of scientism, unless,
following Popper, we mean by it 'the aping of what is widely
mistaken for the method of science'.(8) In a sense, the opinion
expressed in this book is at variance with the anti-scientist
criticism in sociology, and that not only because of the said
thesis on the 'patterns of natural science' being drawn from
natural science. In my opinion, Popper is right when he says
that 'labouring of the difference between science and the
humanities has long been a fashion, and has become a bore,'(9)
with the proviso that in the last decades that fashion was more
kind to papers on the unity of science, and that papers of that
type have proved even more boring. Anti-scientist criticism
joined the discussion in accordance with the principles set by
its opponents, thus attacking the real danger to the social
sciences, but not in the quarter where that danger really
originates. From the anti-scientist point of view, it is the
methodological autonomy of the social sciences, that is at stake
in that discussion, or, to put it more precisely, the development
of those disciplines in accordance with their own logic of development, without patterns and limitations imposed from the
outside. The fact is, that those patterns and limitations are
imposed in the name of natural science, which is, however, not
to say that that is their real origin. Yet the anti-scientist
defence of the methodological autonomy of the social sciences
made a point of looking for methodological differences between
natural science and the social sciences. In this way, in reply to
the doctrine of the methodological unity of science we have the
doctrine of methodological differences between natural science
and the social sciences. Note that both doctrines are of necessity
based on the very dubious assumption that (all or some part of)
scientific production is marked by definite and unchanging
methodological properties. Furthermore, these two doctrines
reveal a far-reaching similarity when it comes to the crux of
the matter. The doctrine of differences, if we may call it so,
being defence-oriented, was in its basic outline determined by
its opponents: the search for specific methodological features of
the social science had its origins in the picture of uniform
science, suggested by the opponents. In such a search, the
picture of the social sciences is limited in advance by those
problems which emerge in connection with the description of
the methodological properties of natural science.

Defence against patterns imposed from the outside has

naturally resulted in the search for specific methodological features of the social sciences being accompanied by a programmatic - not to say reflex-based - dislike of any borrowing from natural science. This can be illustrated by the attitude toward attempts to introduce certain elements of mathematics into the social sciences. Those attempts are still in the early stage of often naive experiments, but works by James Coleman for instance, show that an author who has a good feeling for social issues, can turn such experiments into a serious research project.(10) How far such undertakings can penetrate social problems is a different matter.

In my opinion, reflection on the methodological foundations of the social sciences must make itself independent of reflection on the methodological foundations of natural science. It must make itself independent not in the sense of not admitting any exchange of ideas, but in the sense of carrying on the search on its own, on the basis of an analysis of its own subject matter, i.e., the social sciences, and not on the basis of any methodological model of natural science. The conception of the Geisteswissenschaften, for all its many advantages, is defective in the sense of being logically (and of course, historically as well) derived from the concept of the Naturwissenschaften and offers a profound picture of the humanities, but a picture which - however paradoxical this may sound - is limited by the then current ideas of the methodological properties of natural science. Ossowski's reflections on the methodological properties of the social sciences would have probably followed a different and more fruitful course if they had not been reflections on the peculiarities of the social sciences. The problem of values in the social sciences has suffered by being closely linked to the contrast between description and evaluation. Many other examples could be cited. Generally speaking, the defensive attitude of those who understood that the social sciences could not be reduced to natural science at its lower level of development has not been conducive to far-reaching reflections on the foundations of the social sciences.

Anthony Giddens says that 'the word "positivist", like the word "bourgeois", has become more of a derogatory epithet than a useful descriptive concept, and consequently has been largely stripped of whatever agreed meaning it may once have had.'(11) I do not think that if a given term acquires the properties of an epithet it thereby loses its descriptive value. In any case, the adjective 'positivist' has retained its descriptive value even though it does often occur (but very rarely exclusively) as an epithet. But in sociology there has been a certain conceptual transformation which discourages use of that term in connection with the issues discussed here. That transformation consists in treating the secondary generic features as the primary ones. This has resulted in a far-reaching trivialization of the concept of positivist sociology. As Christopher Bryant noted, 'pre-

occupation with counting, measuring, scaling, with operations
and indices, with survey, statistical calculations and mathematical models, indeed with anything suggestive of "hard"
science, is often referred to as "positivism".'(12) If we add
such identifying marks of 'positivism' as clumsiness in the use
of language and a striking lack of humanistic culture, we have
practically the full 'positivist' syndrome in sociology. This is
why I avoid using the term positivist sociology in referring to
the main trend in sociology which is being criticized here; I
likewise avoid calling my own standpoint antipositivist. It is
a fact that the trend criticized in this book amply reveals all
the characteristics listed above. One could go even further and
point to the specific subculture of academic life produced by
that trend, its bureaucratic ethos, vast connections with non-
academic institutions, etc. But the essence of that trend should
not be interpreted in such terms, in any case not when the
methodological foundations of sociology are being discussed.
Empiricist sociology is a product of positivist thinking in sociology, which during the last fifty years came to permeate that
discipline as a result of the methodological doctrine of logical
positivism. That thinking favours limitations of general humanistic knowledge, focusing of attention on specific research
instruments, assigning fetish-like importance to linguistic correlates of rigorous thinking (often at the cost of rigorous thinking itself), tendencies toward 'shallow empiricism', etc., but
these and other correlates of positivist thought in sociology
cannot be identified with that thought as such; after all they
are not inseparably linked to it.

The impact of the logical-positivist philosophy of science upon
sociology was already quite visible in the 1930s. The 1940s and
1950s marked the turning point, that is, the period in which the
logical-positivist methodological doctrine, after appropriate
adaptation measures which consisted, among other things, in
its vulgarization, became the methodological foundation of the
main trend of sociological research. It has not lost that status,
as I shall try to demonstrate, even to this day, despite far-
reaching changes in the philosophy of science and growing criticism within sociology (which criticism, unfortunately, is often
verbal or based on misunderstanding).
    The success of logical positivism in the social sciences in
general and in sociology in particular, has been something extraordinary. Note that what we have here is not the adaptation
of some methodological ideas or other, but the adaptation of an
entire system of ideas at variance with sociological tradition and
with the research experience of sociologists. Furthermore,
that system of ideas, shaped without any connection with other
problems and in a different intellectual climate, soon won the
status of the methodological foundation of almost the whole of
sociology. This led to the collapse of a fairly rich trend in
humanistic thinking; in the 1950s, sociology was already an

eclectic conglomerate of traditional sociological thought and scientific output of a new type, more and more dominant in academic life, based on new assumptions, and contributing to a quite different scholarly tradition.

This fact can be interpreted in various ways, but every interpretation must take two factors into account. First, the philosophy of science, with its implicit formula for what is scientific, opened new vistas to the social sciences of equating their status with that of natural science; second, all this took place in a favourable intellectual climate, at a time when it seemed that science, 'in the process of conquering new areas,' was just on the eve of conquering the sphere of socio-psychological and cultural phenomena.

These, so to say climatic, conditions explain the otherwise astounding ineffectiveness of that criticism which pointed to both the inadequacy of the proposed methodological patterns and the small value of the resulting scientific output. These conditions probably also account for the practically unlimited opportunities for simplifying and vulgarizing the ideas of logical positivism in the process of their adaptation. These matters have so far passed unnoticed, and yet it was just owing to those simplifications and vulgarizations, that the methodological conceptions of logical positivism were assimilated so readily and have taken such deep roots. In fact, those methodological foundations of 'empirical sociology' which are to be found in handbooks and which function in research practice form a peculiar mixture of the popular version of logical positivism and sociological common sense. This is probably why they meet with kindly silence on the part of philosophers of science, and with acceptance on the part of most sociologists. But they are very susceptible to critical arguments from the point of view of both philosophers of science and sociologists. A compromise between what is popularly held to be scientific and what partially agrees with sociological sense, must sooner or later lose its attractiveness.

It would be a simplification to treat 'empirical sociology' as a simple amplification of the methodological doctrine of logical positivism. The modification to which those methodological ideas were subject in the process of their adaptation has just been mentioned. But even more essential is the fact that 'empirical sociology' has, in fact, a very complex genealogy: as has been suggested by Lazarsfeld,(13) its origins can be seen in European social statistics and simple surveys in the nineteenth and early twentieth centuries (Quételet, Le Play, Booth, etc.), in French positivism, in the empirical studies conducted by the Chicago School, in behaviourism, and in at least several other events in the history of the social sciences (some people even associate the beginnings of that trend with the studies of Thomas and Znaniecki, which, however, is a complete misunderstanding). The rise and development of 'empirical sociology'

can also be treated as a spontaneous reaction of researchers
against the exuberant speculations of armchair thinkers (this
interpretation was at one time quite a stereotype), and as an
equally spontaneous following of the paths of physics and other
sciences (which assumes that sociologists are able to retrace
those paths). Each of these genealogical versions is confirmed
to some extent, but this does not change the fact that the doctrine of logical positivism was decisive for the emergence and
evolution of 'empirical sociology'. This means not only a direct
influence upon the research practice of empirically-minded
sociologists, but also the methodological and epistemological substantiation of that practice. 'Empirical sociology' as an intellectual undertaking, its academic status and aspirations, all this
can wholly be justified only by reference to logical-positivist
methodology from the period of its expansive growth, that is,
from the period which preceded the complications so well described by Hempel, in 'The Theoretician's Dilemma'. Without that
methodological screen 'empirical sociology' as an intellectual
undertaking loses its reason for being (this does not apply to
individual research projects and publications).

Such being the case, the critique of logical-positivist methodology, and a fortiori of the philosophy of science as a scientific
undertaking, indirectly aims at the foundations of 'empirical
sociology'. This is the reason, though not the only one, why
the next two chapters are a tentative critical appraisal of the
philosophy of science. It is in two senses an appraisal by a
sociologist: one point is to bring out those weaknesses of the
philosophy of science which most strongly reflect upon the
evaluation of the present-day situation in sociology; another
point is to view the philosophy of science from the theoretical
background of sociology. These matters are discussed mainly
in chapter 3; chapter 2 in a way introduces chapter 3 by dismissing - it is to be hoped - the objection that what is claimed
in chapter 3 does not comply with what is now taking place in the
philosophy of science especially in its historical trend.

A comprehensive analysis of the impact of the philosophy of
science upon sociology would by far exceed the scope of the
present book. It would require a number of various separate
studies, from a very general reflection to detailed analyses of
the various methodological solutions and their consequences.
Connections between research practice and a definite methodological doctrine are rarely striking ones: the links are, as a
rule, indirect and hidden, so that subtle interpretative analyses
are required to bring them to light. The same applies to the
case we are interested in. For instance, the social indicators
movement is apparently quite independent of any general
methodological doctrine. Rather, it is an activity in the area
of technical operations based on better or worse factual assumptions. Its connections with the doctrine of logical positivism is
quite questionable in advance, because the study of social

indicators is of necessity at variance with the requirement that scientific activity be dissociated from valuation. And yet, as the present writer holds, it can be proved that that under-taking is a misunderstanding if it is not based on certain strong assumptions that have their roots in logical-positivist methodo-logy. One could also try to reveal the historical links between the social indicators movement and logical-positivist conceptions (which does not contradict the fairly obvious claim that this movement was historically connected with social statistics).

I neither intend, nor am in a position to enlarge upon that subject here. This example is only used to illustrate the type of problems that deserve to be taken up, especially insofar as - which can be claimed in advance - the thesis of connections between such cases and logical positivism is strongly contro-versial. For obvious reasons, such analyses would have to be the subject matter of separate publications. In this book I have concentrated on matters which are of primary importance for the problems to be discussed; I have also selected my examples so as to show the cases that are the most telling from the point of view of what is claimed here, and at the same time the most representative in the sense of being associated with great names and with great events in science. To put it briefly, I try to show the most visible symptoms of the impact of the philosophy of science upon sociology, leaving more debatable issues for separate treatment.

Contrary to common stereotypes, I shall not analyze any statements concerning value-free sociology and the correspond-ing aspects of research practice. It is a fact that the descrip-tive purity of science is a dogma in 'empirical sociology', and that that dogma appears there in the context of the system of dogmas of logical-positivist methodology. Thus, Robert A. Dahl, for instance, says that science has as its proper subject matter those 'events which may be observed with the senses and their extension' and for that reason dismisses all valuation from the sphere of the social sciences.(14)

But, on the other hand, as has been pertinently noted by Abraham Kaplan, 'the position on value judgements taken by logical empiricism is not required either by its logic or its empiricism,'(15) which is to say that it is not a necessary appen-dage to the doctrine proper. In other words, the attitude toward value judgments ('a command in a misleading grammatical form' - Rudolf Carnap) was that of logical positivists rather than of logical positivism. It was, in fact, a manifestation of a broader system of opinions that functioned independently of logical positivism. Even more, that system of opinions on values in science, has a long and rich tradition in the social sciences in general and in sociology in particular (a richer one than it has in the philosophy of science). Such being the case, the inter-pretation of statements concerning value-free sociology and the related procedures, as an effect of the impact of the logical-positivist doctrine upon sociology, goes too far. Furthermore,

it is not worth reanalyzing issues which have already been
given exceptionally thorough treatment in the literature of the
subject. This is why that whole group of problems is left
outside the scope of the present book.

This book is intended neither to offer an orderly treatment of
the problems raised above nor, a fortiori, to give a complete
lecture on the subject. It hopes to bring out and articulate
some sore points in present-day sociology, namely those which
are due to its logical-positivist heritage. They have a common
denominator in their lack of methodological tolerance and a
single trend's claim of the sole right to being scientific. It
is important to examine on what grounds that claim is based.
   As has been said, the problems discussed here, do not make
for an impartial weighing of the pros and cons. Furthermore,
I firmly believe that the interest of sociology now requires
looking for and elaborating arguments that would refute the
methodological foundations of 'empirical sociology'. In accordance
with that conviction, I formulate arguments critical of the main
trend in sociology and try to substantiate them. In this sense,
but I hope only in this sense, this book is partial. Such critical
activity, provided it is honest, can cite very respected models
in the past. Naturally, I do not claim to vie with them as to
respectability, and what has been said just now is only meant
to locate this book in the vast spectrum of scientific criticism.
If we agree with Popper that criticism is the motive power of
the development of science, then we have to state that criticism
is a necessary, though not a sufficient, condition for an intellectual revival of sociology.
   This book is not, in any case, to be treated as a historical
study. Historical data form only the substratum on which the
various methodological and theoretical issues related to the
present-day state of sociology are discussed. This is why only
selected examples are drawn from the history of sociology, and
events of now purely historical significance are disregarded.
This applies, for instance, to Lundberg's sociological work.
Lundberg is rightly held to have been an arch-positivist in
sociology, but his radical social behaviourism, operationalism,
and naive physicalism make his opinions like a prototype of
logical-positivism and account for their now being universally
treated as anachronistic. It is a separate issue that, by treating Lundberg's views as the model of logical positivism, people
sometimes construct such a conception of logical-positivist
sociology that compared to it the advanced logical positivism of
the 1950s and 1960s looks like anti-positivism.(16)
   It would be worth while analyzing present-day Polish sociology
from the point of view of these problems. This would be particularly interesting because the impact of the Lvov-Warsaw
School, the defensive attitude toward the dogmatic version of
Marxism and the great prestige of science in this country - to
mention only the most important factors - made Polish sociology

very susceptible to logical positivism. But the manner in which
discussion is carried on in Polish sociological circles, and the
probable political repercussion of a public discussion of the
issue, discourage one from taking up that otherwise fascinating
subject.(17)

2 The history-oriented trend in the philosophy of science: breakthrough or continuation?

The now current belief is that in the 1960s the philosophy of science underwent a profound crisis, due mainly to Kuhn and other representatives of the 'historical schism'. The effects of that crisis are usually summed up thus:

(i)   the practically complete collapse of the classical methodological doctrine of logical positivism;
(ii)   a strong undermining of Popper's methodology;
(iii) a questioning of the raison d'être of the philosophy of science in the form it had so far, i.e., as the logic of science.

Without questioning the fact that the crisis has been profound and that its effects are like those listed above(1) it is possible to interpret the present situation in various ways. This chapter is a tentative formulation of certain arguments intended to minimize the importance of 'the turn toward history' and in particular to claim that we are dealing with a breakthrough which mainly affects the style of philosophical work and which, while it has far-reaching consequences, does not undermine the foundations of the traditional philosophy of science.

Let us begin with the fact that this turn toward history can be described in general terms as a turn toward history written by philosophers, and not by historians. This is due to a particular way of pursuing the history of science, to which Ernam McMullin refers thus:

> What bothers historians of science about this is that it often seems to them to be masquerading as history; it makes use of the great scientists of the past as lay figures in what seems to be a historical analysis but really is not. They are manipulated to make a philosophical point which, however valid it may be in itself, was really not theirs, or at least is not really shown (using the proper methods of the historian) to have been theirs. Though names of scientists and general references to their works may dot the narrative, what is really going on (in the sense of where the basic evidence for assertions ultimately lies) is not history.(2)

Let us examine the issue more closely by taking as our starting point Imre Lakatos's reflections on the history of science.(3)

He begins by paraphrasing Kant: 'Philosophy of science without history of science is empty; History of science without philosophy of science is blind.' The first clause substantiates the turn toward history; the second makes one expect that Lakatos, being in favour of links between philosophy and history, will suggest an assignment of parts which would make philosophy the sphere of thought, and history, the sphere of 'facts.'

In Lakatos's opinion, pursuing the history of science consists in interpreting science's past in terms of the methodological system adopted:

> Some historians look for the discovery of hard facts, inductive generalizations, others for bold theories and crucial negative experiments, yet others for great simplifications or for progressive and degenerating problem shifts.(4)

Mentioning these interests of historians of science rather than others is linked to Lakatos's view of present-day methodological doctrines. He singles out inductivism, conventionalism, falsificationism, and his own doctrine, namely 'the methodology of scientific research programs.' Accordingly, he classes historians of science as inductivists, conventionalists, etc. The question arises, of what the difference is between a historian of science, interpreted in this way and a philosopher of science, if the latter has the duty, imposed upon him by Lakatos, of studying historical data as the empirical basis of his conceptions, and the former makes use of those conceptions in studying historical data. What, for example, is the difference between an inductivist philosopher and an inductivist historian if:

> The inductivist historian recognizes only two sorts of *genuine scientific discoveries*: *hard factual propositions* and *generalizations*. These and only these constitute the backbone of his *internal history*. When writing history, he looks out for them - finding them is quite a problem. Only when he finds them, can he start the construction of his beautiful pyramids. Revolutions consist in unmasking (irrational) errors which then are exiled from the history of science into the history of pseudoscience, into the history of mere beliefs: genuine scientific process starts with the latest scientific revolution in any given field.(5)

The answer lies in the phrase 'when writing history'. When one 'is writing history', one is a historian, when one 'is writing philosophy', one is a philosopher; the division is purely functional. There are no inductivist historians nor inductivist philosophers, there are just inductivists, who at one moment are inclined to show science's past, and at another moment to explain the methodological model of science, but who always represent the same system of views which is termed inductivism.

Furthermore, that system of views, whatever we might say
about it, is a system of philosophical views. It should be stated
that 'a historian of science is a philosopher of science interested
in the past'. Likewise, accordingly to Lakatos, history of science
becomes philosophy of science in a historical account.
    Lakatos treats his conception of the history of science as
a rational reconstruction of research practice, i.e., as a
description not so much of what the historian of science really
does, as of what he would do if he acted rationally. In such
a case, in order to substantiate one's views one must refer,
directly or indirectly, to some theory of rationality. Lakatos's
argumentation, while it goes in that direction, is extremely
simple: he says that one cannot engage in the history of science
without some conception of science that would delimit the class
of facts under consideration. Every historian of science whether
consciously or not, does use some conception of science. Such
a conception may be very vague or fuzzy, but that is a matter
of that historian's working techniques. If so, then a history of
science without a philosophy of science is unthinkable. The
historian must explicitly or implicitly refer to some normative
methodological system which determines his research problems
or, more precisely speaking, the internal history of science,
i.e., determines what belongs to science and what lies outside
it, and will eventually become - insofar as it is related to its
internal history - the external history of science. Hence,

> Whatever problems the historian of science wishes to solve,
> he has to reconstruct the relevant section of the growth of
> objective scientific knowledge, that is the relevant section
> of internal history. As has been shown, what constitutes
> for him internal history, depends on his philosophy, whether
> he is aware of this fact or not.(6)

This necessary connection between the internal history of
science and the philosophy of science is a basic element of
Lakatos's argumentation. He realizes that on this point his
conception moves away from the research practice of professional
historians of science (and the analogous statement holds, e.g.,
for the sociology of science). As is usual in such cases, he
sees in this fact an argument to be used against this research
practice:

> Long texts have been devoted to the problem of whether,
> and if so, why, the emergence of science was a purely
> European affair; but such an investigation is bound to
> remain a piece of confused rambling until one clearly defines
> 'science' according to some normative philosophy of sci-
> ence.(7)

This concludes that part of Lakatos's argumentation which is
the most important one from our point of view. The essential

fact is that Lakatos precludes the possibility of pursuing the history of science independently of the philosophy of science, and thus questions the sense of all that 'turn toward history' in favor of which he has declared himself. It is very unclear how the history of science, seen through a normative methodological doctrine, could bring about a change in the doctrine itself. As has been noted by Kuhn, a philosopher who makes use of Lakatos's conception can draw from the history of science only such conclusions concerning methodology as he has previously put into it.(8)

But Lakatos goes further and brings his conception to the point at which it completely ceases to correspond to research practice in the history of science and at which using that term becomes merely a façon de parler. Here is the passage in question:

> Thus in constructing internal history the historian will be highly selective: he will omit everything that is irrational in the light of his rationality theory. But this normative selection still does not add up to a fully fledged rational reconstruction. For instance, Prout never articulated the 'Proutian programme': the Proutian programme is not Prout's programme. *It is not only the ('internal') success or the ('internal') defeat of a programme which can only be judged with hindsight: it is frequently also its content.* Internal history is not just a *selection* of methodologically interpreted facts: it may be, on occasion, their *radically improved version.* One may illustrate this using the Bohrian programme. Bohr, in 1913, may not have even thought of the possibility of electron spin. He had more than enough on his hands without the spin. Nevertheless, the historian, describing with hindsight the Bohrian programme, should include electron spin in it, since electron spin fits naturally in the original outline of the programme. Bohr might have referred to it in 1913. Why Bohr did not do so, is an interesting problem which deserves to be indicated in a footnote. (Such problems might then be solved either internally by pointing to rational reasons in the growth of objective, impersonal knowledge; or externally by pointing to psychological causes in the development of Bohr's personal beliefs.)
>
> One way to indicate discrepancies between history and its rational reconstruction is to relate the internal history *in the text*, and indicate *in the footnote how actual history 'misbehaved'* in the light of its rational reconstruction.(9)

In other words, the historian's basic work, i.e., pursuing the internal history of science, consists in rationally reconstructing the history of science, that is, not so much in describing what in fact did take place (even if seen through the methodological doctrine he has adopted) as in describing what would have occurred if the history of science were a purely rational process, i.e., a process that is in agreement with the methodological theory of rationality, adopted in the doctrine in question. Hence that 'improvement' on historical facts, which does not mean falsifying history, but turns so-called historical truth - which for professional historians is the basic goal of their research - into a subject matter of footnotes. As has been noted by McMullin, those footnotes often deny what is written in the main body of the text. For instance, in the main body of his text, Lakatos claims that Prout *'knew very well that anomalies abounded* but said that these arose because chemical substances as they ordinarily occurred were impure, that is, the relevant experiment techniques of the time were unreliable' but informs in the footnote: 'Alas, all this is rational recon-struction rather than actual history. Prout denies the existence of any anomalies.'(10)

Quite understandably, a 'history' which by assumption admits 'improvements' on facts is hardly acceptable even to the most tolerant historians.(11) Note, however, that - perhaps even without noticing it - we have found ourselves in the sphere of verbal controversies: the issue boils down to the fact that Lakatos, evidently contravening good terminological manners, puts forth his idea under the name 'history of science' whereas he should put it forth as 'a logical reconstruction', or rather 'a rational reconstruction', of science. If his proposals were confined to the demand that science's past be studied in the light of normative methodology, we would be dealing with a highly debatable proposal, but nevertheless a proposal on how the history of science should be written. But if it turns out in the end that the historian carries out a rational reconstruc-tion (possibly completing it with anecdotal data), i.e., does what has been suggested, by Popper for instance (who, how-ever, did not call that history of science), then Lakatos's reflections must be treated as a rather extravagant proposal concerning the use of the term 'history of science'.

This is not to say in the least that Lakatos's conception of history of science is due to a confusion of ideas. But it is a result of conceptual operations intended to link, by an excellent argumentation, the old conception of the logical reconstruction of science with the now ennobling term 'history of science'. We seem to have a case here of defending an endangered theory by an appropriate (neutralizing) interpretation of the premises from which the danger comes.

Unlike Lakatos, Paul Feyerabend is inclined to improve the philosophy of science on the basis of history of science rather than conversely. He is in fact one of the most consistent critics of present-day philosophy of science, and especially of the Popper school (being himself connected with that school), and his arguments are based on an analysis of historical data. He claims that these data allow us to state that the contemporary philosophy of science offers a wrong picture of science and

suggests erroneous rules of research procedure. To make
matters worse, it follows the wrong track by searching for
nonexistent things, i.e., constant methodological rules.(12)
His analysis of historical data makes him formulate an 'anarchistic', as he calls it, methodology of science, whose leading
principle is 'anything goes.'(13)

Feyerabend's historical research takes on the form of case
studies. This is an essential fact to which he assigns much
importance:

> Lakatos has convinced me [he means Proofs and Refutations,
> 'The British Journal for the Philosophy of Science', 14,
> 1963-4] that Kuhn's approach is still too abstract and that
> there is only one way to acquire the needed information:
> case studies. But case studies - and this seems to be the
> main point of this essay - do not need to degenerate into
> gossip (as does much specialist history), but can be truly
> philosophical, being guided by abstract principles and
> giving rise to them in turn.(14)

What is gossip to a philosopher may be essential information
to a sociologist (historian, anthropologist, etc.). And conversely: a 'truly philosophical' case study may be of purely
anecdotal value for a sociologist. Feyerabend suggests a false
alternative: either philosophy or gossip. He probably did not
mean that; in any case, it does not fit well with his methodological views. But it fits extremely well with his own way of
pursuing history of science. In other words, whatever Feyerabend qua methodologist of history thinks about the alternative:
to gossip or to philosophize, Feyerabend qua historian of
science sees only these two possibilities. As a result, he uses
historical cases to philosophize; his case studies are interpretations of philosophical conceptions,(15) and the requirements
of writing history are subordinated to the needs of philosophical
argumentation. As McMullin has demonstrated by analyzing in
detail what Feyerabend wrote on Galileo's cosmogony, Feyerabend does not comply with the requirements of the techniques
of writing history, selects facts in a way which is not substantiated by historical data, formulates generalizations based
on atypical cases, interprets events without taking into account
the historical context, and finally, distorts the sense of
historical events by presenting them in a dramatized form.
McMullin's conclusion is unambiguous:

> Despite appearances, therefore, Feyerabend's PS (Philo-
> sophy of Science) in this paper ('Problems of Empiricism',
> II., op. cit.) does not rest upon HS (history of science).
> He brings a prior notion of rational inquiry to bear upon
> the history of science with a view to finding there some
> support for his view. But the way in which he does this
> forces one to say that his PS is *not* grounded in history;

whatever support it has, it must draw from elsewhere. It is thus, a PSE (external philosophy of science, that is a philosophy of science whose 'warrant is not drawn from an inspection of the procedures actually followed by scientists'), not a PSI (internal philosophy of science), and it is a PSE of peculiarly risky sort in that by purporting to be a PSI, it is effectively exempted from exploring its real warrant.(16)

Thomas S. Kuhn is certainly engaged in valuable and original studies in the history of science. Historical data are for him neither - as in Lakatos's case - the subject matter of 'rational reconstruction', nor - as in Feyerabend's case - the level on which the philosophy of science drawn from elsewhere is explained; nor are they in any other way reduced to the role of an auxiliary factor. But, on the other hand, Kuhn's principal works have not gone outside that which Gouldner would have called background assumptions of the philosophy of science,(17) i.e., outside the framework of an unexplicated system of beliefs which determine the cognitive perspective of the philosophy of science, make the representatives of that discipline concern themselves with problems of a definite kind, which make them structure facts in a definite way, etc.

It is a common knowledge that answers depend primarily on questions. The questions which Kuhn poses and to which he seeks answers in historical data have been shaped by the philosophy of science and are closely connected with the problems investigated by the Popper school.(18) Whatever then the merits of Kuhn's historical analyses, their general sense has been determined by philosophy rather than historiography. Kuhn accordingly does not so much introduce the historical approach to the philosophy of science as show the significance of the study of science's past for philosophical reflections on science. The meaning of this claim will become clearer later.

Let us pose the next question: why is it just the history of science that has become a point of interest for philosophers of science? Kuhn claims a role for history; Amsterdamski(19) calls for making the philosophy of science history-oriented; Lakatos's already quoted statement, that philosophy of science without history of science is empty, has almost become a slogan. What then does history of science add to philosophy of science? What does it offer, in fact or fiction, to make philosophical reflections on science seek its support?

For Lakatos, history of science is the sphere of the methodological critique of the theory of science (and hence of rational reconstructions of the history of science):

Historiographical falsification of inductivism ... was initiated already by Duhem and continued by Popper and Agassi. Historiographical criticisms of (naive) falsificationism have been offered by Polanyi, Kuhn, Feyerabend and Holton. The

most important historiographical criticism of conventionalism
is to be found in Kuhn's ... masterpiece on the Copernican
revolution. The upshot of these criticisms is that all these
rational reconstructions of history force history of science
onto the Procrustean bed of their hypocritical morality, thus
creating fancy histories, which hinge on mythical 'inductive
bases', 'valid inductive generalizations', 'crucial experiments',
'great revolutionary simplifications', etc.(20)

To claim this, one has to assume that the methodological theory
of science is subject to confrontation with facts and that among
those facts, with which the methodologist must reckon, is
science's past. It is known, that this was asserted by Kuhn in
'The Structure of Scientific Revolutions', where he stressed
the second part of the thesis rather than the first:

Having been weaned intellectually on these distinctions [the
context of discovery vs. the context of justification] and
others like them, I could scarcely be more aware of their
import and force. For many years I took them to be about
the nature of knowledge, and I still suppose that, approp-
riately recast, they have something important to tell us.
Yet my attempts to apply them, even *grosso modo*, to the
actual situations in which knowledge is gained, accepted,
and assimilated have made them seem extraordinarily prob-
lematic. Rather than being elementary logical or methodological distinctions, which would thus be prior to the analysis
of scientific knowledge, they now seem integral parts of a
traditional set of substantive answers to the very questions
upon which they have been deployed. That circularity does
not at all invalidate them. But it does make them parts of
a theory and, by doing so, *subjects them to the same scru-
tiny regularly applied to theories in other fields*. If they
are to have more than pure abstraction as their content,
then that content must be discovered by observing them in
application to the data they are meant to elucidate. *How
could history of science fail to be a source of phenomena
to which theories about knowledge may legitimately be asked
to apply?*(21)

Putting Lakatos and Kuhn together is not accidental in this
context, for it illustrates the fact that Kuhn carried the day
very easily by advancing the thesis formulated above. In some
cases (as in that of Lakatos) the victory was not complete
(Lakatos indirectly interpreted that thesis in his own way, at
variance with Kuhn's intentions, on how the methodologist is
to put his respect for history of science into effect), but in
any case his success was spectacular, at least in the sphere of
declarations and the phraseology used, and that also in the
milieu at which his criticism was aimed.
  In my opinion, that fact can be explained as follows: by

advancing his thesis Kuhn did not infringe any current principle of philosophy of science. His criticism was aimed at the traditional way of pursuing philosophy of science, inspired by logical positivism, but not at the fundamental assumptions of that discipline. For these assumptions are not only not in contradiction with the thesis in question – they actually imply it.

The conception of methodology advanced by logical positivists assumed that the methodological doctrine is, all else aside, a reconstruction of the methodological properties of science. At first, the methodological reconstruction of science was used to fight metaphysics and developed to serve that purpose: the point was to bring out those properties of science which make it differ from metaphysics. As interest in 'the problem of metaphysics' declined, methodological research turn toward reconstructing the basic methodological properties of science. The main objective now was not to draw the demarcation line between science and that form of non-science which is metaphysics, but to offer a uniform picture of the essence of science, as Kemeny put it. In fact, that picture of the essence of science – like the demarcation criterion applied earlier – originated rather from a certain system (or, more cautiously, set) of assumptions pertaining to rational cognition than from an analysis of what really takes place in science, and the arguments used referred mainly to logical reasons, leaving empirical data the role of illustrations. Hence the objection that the methodology of logical positivists resembles exercises in logic is – if we disregard the expressive aspect of this formulation – largely to the point, but

(i) it refers to the implementation of the research program, and not to programmatic assumption;
(ii) the exercises were certainly not in pure logic, in view of the fact that the methodological doctrine of logical empiricism was undergoing far-reaching modifications, and each modification marked an endeavor to eliminate evident discrepancies between the doctrine and the facts which that doctrine was meant to describe.

In Popper's work all this looks somewhat different: methodology is by assumption primarily a normative model of science. This was one of the many points on which Popper disagreed with the Vienna Circle. During the discussions on demarcation Popper described his standpoint, which opposed that of the Vienna Circle, thus:

Positivists usually interpret the problem of demarcation in a naturalistic way; they interpret it as if it were a problem of natural science. Instead of taking it as their task to propose a suitable convention, they believe they have to discover a difference, existing in the nature of things, as it were, between empirical science on the one hand and metaphysics on the other.(22)

And further:

> My criterion of demarcation will ... have to be regarded as
> *a proposal for an agreement or convention*. As to the suit-
> ability of any such convention opinions may differ; and a
> reasonable discussion of these questions is only possible
> between parties having some purpose in common. The choice
> of that purpose must, of course, be ultimately a matter of
> decision, going beyond rational argument.(23)

Of course, Popper at no point abstained from arguing rationally in favor of his model of science. Truth was the value he
stressed most, but he linked that value, in doing which he followed current ideas, to a number of social values of primary
importance, so that the limits of rational discussion did not have
to be reached. Be that as it may, Popper's model of science is,
by assumption, based on a certain conception of rational procedure which strives for true knowledge. In this sense, he
primarily answers the question what properties science should
have. But, on the other hand, that model is consciously, and
by assumption, constructed so as to, first, cover the accepted
achievements of science, primarily physics, in which Popper
always saw the paragon of science ('the most complete realization to date of what I call "empirical science"'), and second,
to exclude that which in the opinion of Popper himself and those
circles of scientists which were close to him was considered
negative examples. As Lakatos writes (referring to Popper's
reminiscences), Popper arrived at his criterion of demarcation
by starting from an analysis of selected cases in the recent
history of science, with the proviso that he already had an
earlier formed opinion on those cases:

> He thought, like the best scientists of his time, that Newton's
> theory, although refuted, was a wonderful scientific achieve-
> ment; that Einstein's theory was still better, and that astro-
> logy, Freudianism, and twentieth-century Marxism were
> pseudo-scientific. His problem was to find a definition of
> science which yielded these 'basic judgements' concerning
> particular theories; and he offered a novel solution.(24)

Thus one could say: even though Popper's methodological
model was conceived as an explication of the principles of
rational knowledge, it is also, by assumption, a description of
science in its best version, i.e., 'Great Science', as Popper
sometimes used to call it. Obviously, the limits of that Great
Science are not quite independent of the adopted theory of
rationality, but have been determined in principle by social
factors (in the sense of complying with current evaluations)
and in this sense they logically precede the theory of rationality. There is no doubt that for Popper it is a dogma that certain facts belong to science, and this codetermines his concept

of science. Popper was always inclined to underestimate even
the most widely accepted achievements of the social sciences.
His position was consistent: those achievements were not on the
socially determined list of Great Science, and hence they
deserved recognition only under the condition that they agreed
with the methodological model, which condition they obviously
did not meet.(25) On the other hand, Popper could not underestimate the recognized achievements of physics, because that
would be at variance with his basic assumptions: those achievements by assumption deserved being included in Great Science,
and the methodological model merely sanctioned the fact.

If Kuhn's works are seen against this background, it is
apparent that they are in fact revolutionary, not in the sense
(often claimed for them) of having introduced a new epistemological paradigm into the philosophy of science, but rather in the
sense of introducing, on the basis of current assumptions of
the philosophy of science, a new model of research procedure,
a model with very fruitful consequences. Contrary to established
research practice, Kuhn began just systematically using the
empirical material. This must have resulted in an upheaval in
the philosophy of science as a whole: on the one hand, this
procedure was in agreement with the theoretical foundations of
that discipline and hence was difficult to dismiss, and on the
other hand, it yielded results most unfavorable to the most
respectable methodological doctrines.

By pleading for a role of history Kuhn thus postulated a
confrontation of methodological doctrines with the empirical data
in the history of science, which requirement complied with the
current assumptions. An analogous role is assigned to the history of science by the entire history-oriented trend in the
philosophy of science: the turn toward history means a turn
toward facts or, to put it more precisely, toward those facts
which have so far been least noticed by philosophers of science.

Note that Kuhn and other representatives of the history-
oriented trend in the philosophy of science in their studies
cover a very limited area of historical facts, namely a certain
characteristic period in the history of European physics. That
such a limited observation field naturally provides limited
opportunities for drawing conclusions is clearly disregarded by
them. And yet this is only apparently a methodological error
because we are dealing here with something else, namely with
an assumption which narrows down the scope of the history of
science to the history of European natural science, with emphasis on physics and related disciplines. Kuhn proceeds as if he
were outlining the scope of events covered by history of science
by following the successive antecedents of the most advanced
spheres of present-day scientific thought back to their sources,
which, for him, are never located in any remote past: chemistry
and the theory of electricity go back to the mid-eighteenth
century, genetics, to the mid-nineteenth century, and the old

Greek theories of matter, being in the sphere of philosophy, lie far outside the limits of physics.(26) History of science reaches now further now less far back, according to the given discipline, with the proviso that some disciplines, which Kuhn terms proto-sciences, have not yet entered the phase of truly scientific development at all; this applies to 'many social sciences' among others.

What is the origin of these boundaries and divisions? Is some successive criterion of demarcation involved here, which like the criteria advanced by logical positivists, arbitrally defines the meaning and extension of the concept of science? Undoubtedly so. It is true that Kuhn's theory emerged as a description of the mechanism by which science evolves and not as a basis for distinguishing science from the other fields of intellectual activity, yet in the course of time Kuhn began to treat his theory as a criterion of demarcation, as well. Typically enough, this came about through the impact of criticism: the attempts to avoid criticism led, among other things, to a narrowing down of the concept of science so that any case at variance with the theory would remain outside the field of observation. It was with this end in view that Kuhn, in reply to Popper's criticism, removed the Greek theories of matter from the field of history of science; with this end in view as well, he introduced the concept of proto-science.

The last named concept has proved extremely handy. Kuhn describes proto-sciences as disciplines 'in which practice does generate testable conclusions but which nonetheless resemble philosophy and the arts rather than the established sciences in their developmental patterns.'(27) This obviously makes the criteria of applicability of the term *proto-science* depend on which conception of development of science is adopted. Kuhn is thus in a position to treat cases at variance with his conception of the development of science as being by definition outside the scope of that conception, even if these were cases that under other criteria or traditions could be entirely included within the bounds of history of science. To this he adds the following characteristic vision of the origin of science:

> Each of the currently established sciences has emerged
> from a previously more speculative branch of natural philo-
> sophy, medicine, or the crafts at some relatively well-
> defined period in the past. Other fields will surely experi-
> ence the same transition in the future.(28)

Neither the concept of established sciences nor the distinctive features of that turning point are characterized independently of the theory of development of science advanced by Kuhn, but they occur in the context of that theory immediately after the introduction of the concept of proto-science. Such being the case, the passage quoted above is still another, and extremely suggestively formulated, rule for interpreting those cases which disagree with Kuhn's conception of the development of science; that rule states that the limits of the maturity of a given discipline must be adjusted so as to suit said conception of the development of science; all the cases which disagree with that theory are moved outside the sphere of science proper. (As follows from the second sentence of the rule, such disciplines will move into that sphere in due time, which allows us to say that the cognitive value of Kuhn's theory relates to all disciplines, since it describes the present state of some of them, namely the mature disciplines, and the future state of the others, namely of the proto-sciences.)

Be this as it may, Kuhn uses his own criterion of demarcation, which, while not completely unambiguous, is on many points certainly more restrictive than, e.g., the Popper criterion. I do not think, however, that this fact explains much when it comes to the astonishing - at least for a historian - restriction of the history of science to, broadly speaking, one thread of thought connected with a single culture. The explanation should rather be sought elsewhere, namely in popularly held ideas about science. These ideas underline the entire positivism-oriented trend in the philosophy of science, and have contributed much to the emergence of the history-oriented trend within that philosophy.

As we recall, Popper's point of departure was the commonly held structuring of intellectual activity, in which science par excellence, on the one hand, and pseudo-science, on the other, are singled out with particular clarity. Acceptance of this structuring came to weigh upon Popper's doctrine as a whole, and not on his criterion of demarcation alone, as the structuring lined both the subject matter of research and the general stereotyped picture of that subject matter (if only by providing a list of positive and negative examples). Popper's methodological system, apart from being a fascinating conception of rational cognition, is an extended rationalization of the common sense view of a certain fragment of modern European culture, a rationalization which assigns to that fragment an absolute value by raising it to the rank of a phenomenon common to all men. That which constitutes one of the many forms of rational cognitive activity has been identified with rational activity pure and simple.

Lakatos says that Popper viewed and assessed the various events in the history of science in accordance with the opinions of the best scientists of his time. This is certainly true, especially if we assume that the scientists concerned were the best in the sphere of natural science. But if this is so, it means that what Popper did was to follow the opinions of those people particularly strongly connected with a specific way of thinking, a specific system of values, etc. Their profession made those people view science in a particularly narrow way, assign an absolute value to their own perspective of cognition, notice subtle differences within their own tradition, and disregard the plurality and variety of other intellectual traditions. Further-more, that milieu of best scientists represented current (of course, at an educated level) historical knowledge and common sense view of the world of culture, that is, that cognitive per-spective which makes one see cultural phenomena as they are manifested in the unique context of one's own epoch and its problems. When applied to science, this means seeing it through the prism of current scientific life with its actual achievements, problems, and connections with other spheres of life. In this perspective, the history of science always means merely the genealogy, expanded more or less as the case may be, of cur-rently accepted achievements of science; the rest belongs to other spheres of life or is just the history of errors.

What Popper more or less consciously adopted as his point of departure is a rather automatically accepted basis of all philo-sophical reflection in science derived from the inspiration of positivism. As has been noted by Habermas, logical positivism practically put an end to epistemology by reducing it to the philosophy of science through identifying knowledge with scientific knowledge.(29) This is a pertinent observation, but it has to be further explicated. Logical positivists assigned absolute value not to scientific knowledge in general (whatever this might mean), but to that form of knowledge which they ascribed to present-day natural science, which they identified with science in general. Thus they replaced epistemology not so much by the philosophy of science, as by philosophical reflec-tion on what common sense knowledge treated as science at the time when logical positivism was being shaped.

This legacy of common sense thinking proved an inalienable feature of the philosophy of science inspired by logical positi-vism. It has also been taken over, with all its consequences, by the history-oriented trend in that philosophy. In this respect Kuhn continues the line represented by Carnap, Hempel and Popper. His works, revolutionary as they are in other respects, pertain to practically the same sphere of phenomena, singled out on the same basis.

Now this is a sphere of phenomena which differs from the subject matter of humanistic reflection on science, whether pursued within historical studies, theory of culture, sociology, or social anthropology. Not only is it incomparably narrower:(30) it is located - and this is the essential issue - within the context of another system of opinions, of another cognitive perspective. Generally speaking, it is an element of a different structuring of the world of culture.

It is therefore understandable why for Kuhn and other authors of that group, a turn to history always means a turn to the past of science, to historical facts, etc., but is never a turn to historical theoretical conceptions. But making the philosophy of science truly historical (whatever this might mean) can only be achieved through confrontation with history as an entire intellectual system. Only a confrontation with a thus understood history could essentially effect the direction in which the philosophy of science develops. Confrontation with history interpreted as the past of what common sense knowledge includes within science could, and did, only cause a breakthrough in the methods of the philosophy of science. It is another point that these changes had far-reaching substantive consequences by undermining, among other things, the belief in the existence of a single and changeless form of rational knowledge, which is a preliminary condition for understanding those forms of rational knowledge which have always been outside the observation field of traditional philosophy of science.

3 The philosophy of science in the perspective of the theory of culture

For Lakatos, there is no doubt that history of science without philosophy of science is blind or, to put it more rigorously, impossible, because all historical study of science is logically preceded by a normative philosophy of science which defines what belongs to science, i.e., what is science and what is not. As long as the term 'science' is not clearly defined in accordance with some normative philosophy of science, history of science will be just 'a piece of confused rambling'.(1)

Let us consider analogous cases. Can one study history of religion without a philosophy of religion, history of law without a philosophy of law, history of magic without a philosophy of magic? While adopting an appropriate interpretation of that statement, one can agree that, all historical study of religion must logically be preceded by a concept of religion, but such a concept need not have much to do with philosophy. The claim that history of religion depends upon philosophy of religion and that there is a necessary connection between the 'internal history of religion' and an 'accepted philosophy of religion,' etc., would be an obvious nonsense, and that in the light of research practice as well. Exactly the same can be said about law, magic, art, political thought, etc. Is science in a different position?

The answer to this question sets the modern philosophy of science in opposition to the social sciences and humanities. The former adopts an approach to science which makes one answer in the affirmative. Lakatos's claim is a logical outcome of that approach and in that sense is simply one of the propositions of the philosophy of science regardless of how widely it is really spread in that discipline. The latter disciplines treat science, like religion and magic, in the context of a broadly understood theory of culture, which deprives Lakatos's argumentation, on the dependence of history of science upon philosophy of science, of a fundamental element, namely the claim that there is a necessary connection between the conception of science (and hence the internal history of science) and a normative philosophy of science. The present chapter is concerned with explication of these two opposing approaches. I feel that the nature of what the philosophy of science undertakes is incomprehensible unless seen in the perspective of the theory of culture.

Malinowski began his 'Magic, Science and Religion' thus:

> There are no peoples however primitive without religion and
> magic. Nor are there, it must be added at once, any savage
> races lacking either in the scientific attitude or in science,
> though this lack has been frequently attributed to them.(2)

The substantiation of this statement is based, among other
things, on the well-known distinction between the two spheres
of culture, the sacred and the profane. The former covers (not
only in Malinowski's case) magic and religion, the latter is
identified by him with science: 'In every primitive community,
studied by trustworthy and competent observers, there have
been found two clearly distinguishable domains, the Sacred and
the Profane; in other words, the domain of Magic and that of
Science.'(3)

Thus, in Malinowski's eyes science is in opposition to the
sacred sphere of culture and forms the sphere of rational knowledge, that is, one based on experience and correct inference.(4) He accordingly defines science as 'a body of rules and
conceptions based on experience and derived from it by logical
inference, embodied in material achievements and in a fixed form
of tradition and carried on by some sort of social organization.'(5)

And here is a typical example in favor of the claim that science
is not absent from primitive societies:

> They [the Melanesians whose life Malinowski here examines]
> understand perfectly well that the wider the span of the
> outrigger the greater the stability yet the smaller the resis-
> tance against strain. They can clearly explain why they have
> to give this span a certain traditional width, measured in
> fractions of the length of the dug-out. They can also explain,
> in rudimentary but clearly mechanical terms, how they have
> to behave in a sudden gale, why the outrigger must always
> be on the weather side, why one type of canoe can and the
> other cannot beat. They have, in fact, a whole system of
> the principles of sailing, embodied in a complex and rich
> terminology, traditionally handed on and obeyed as ration-
> ally and consistently as is modern science by modern sailors.
> How could they sail otherwise under eminently dangerous
> conditions in their frail primitive craft?(6)

Malinowski has often been blamed for having used too broad
a concept of science. Too broad relative to what? Now, as a
rule, it turns out to be too broad relative to his critics' conception, such a conception being most often a philosophical
one. For instance, Judith Willer claims that 'Malinowski confused
science with ability to intelligently cope with the environment',
and her argument is as follows.

There are four kinds of knowledge: magical, mystical,

religious, and scientific. This division results from the division
of thought into empirical, rational, and abstractive. Empirical
thought is concerned with connections at the observational level;
rational thought, with those at the theoretical level; abstractive
thought, with those between these two levels. The various
kinds of knowledge are products of the various combinations
of these three kinds of thought.

Magical systems of knowledge are characterized by empirical
thought alone, religion combines abstraction with rationalism,
mysticism makes empirical and abstractive connections, and
the combination of all three types of connection is characteristic of scientific systems.(8)

Science is thus the most complex system of knowledge and
has clear distinctive features.

All thinking which *combines* rational, empirical and abstractive thought is scientific. Neither catalogues of empirical
facts nor rational systems such as mathematics are scientific
thinking by themselves. No system of knowledge is scientific unless it connects the observational and theoretical
levels.(9)

Willer draws a number of conclusions from this, including
one which undermines Malinowski's statement: 'The ability of
"savages" to build ships that float has no more to do with the
understanding of scientific laws than does the effectiveness of
a cook depend upon the understanding of modern chemistry.'(10)
There may be excellent reasons in favour of Willer's interpretation of science, but that is not an argument against other
conceptions of science, especially if these are formulated in
the context of other problems and another tradition.
Malinowski's conception of science is an element of his functional theory of culture and in a sense was necessitated by
that theory. Of course, Malinowski could have defined science
in various ways, but it was a foregone conclusion that

(i) it would be a definition in terms of the theory of culture (and not, e.g., in terms of the logical structure of
the system of knowledge);
(ii) science would be interpreted from the point of view
of the function it performs in the system of culture;
(iii) science - like e.g., law - would prove to be a necessary element of the group's cultural endowment.

Any concept of science that did not meet these conditions
would be of little use in Malinowski's works, if not downright
artificial.
As we know, Malinowski was marked both by the determination
with which he proclaimed his theoretical views and the consis-

tence with which he followed these views when doing research.
He was primarily a field worker, and his lasting achievements
are linked to that role, but he was a researcher with an exceptionally strong theoretical awareness. The fundamental theses
(dogmas, as some of his critics say) of his functional theory of
culture include the following:

(a) Culture is a man's secondary environment, a specific
microcosm in which man organizes his life and thanks to which
he can satisfy his needs.
(b) Culture forms an indivisible whole; its various elements are
interconnected and each of them plays a definite role in the
life of the group.

From these two theses – which obviously functioned in his
system as the most basic assumptions – Malinowski drew a
number of conclusions which he jointly treated as a universal
research method that was practically changeless (and, we have
to admit, little responsive to confrontation with empirical data).
One of these direct conclusions was the thesis that if an element
of culture does exist, then it must play some role in the life
of the group. This claim functioned as a research directive and
an interpretation rule probably to a greater extent than his
other theses did. This was in fact the most conspicuous and
most exploited thesis of the functional theory of culture in the
version adopted by Malinowski and his most faithful disciples.
And it was precisely this thesis which Malinowski used to
explain the existence of such 'irrational' elements of culture as
magic. Whatever one might say about the theoretical values of
this thesis, it proved to be an exceptionally fertile research
directive, at least in Malinowski's own works.
For Malinowski it was a dogma that primitive man is no less
rational than the contemporary European. The principles of the
functional interpretation of primitive cultures thus came to
include the recommendation that all rational aspects of those
cultures be emphasized, which was also closely connected with
the general vision of culture as an instrument of survival.
There was, of course, the problem of the criteria of rationality.
In practical terms, one had to answer the question of what
testifies to the rationality of primitive man. Now Malinowski
solved that problem in accordance with the intellectual conventions of his times and drew patterns of rationality exclusively
from the traditions of European culture. The striving to oppose
the current belief, still lively among anthropologists themselves, that primitive culture is irrational (pre-logical, mystic)
by nature and to put an end to the nineteenth century version
of the division of the world into us and barbarians gave rise
to an undertaking intended to obliterate the difference between
primitive culture (the value of that conceptual category was
still beyond doubt) and European culture. This consisted to
some extent in stressing irrational elements in European culture,

but more often, in accordance with the subject matter of anthropology as understood at that time, in emphasizing the rational
aspects of primitive culture, i.e., those aspects which could be
treated as rational in accordance with the state of European
consciousness in that period. Briefly, when searching for
manifestations of rationality in primitive culture, Malinowski
was looking for traces of that in which the rationality of European culture, measured by patterns current in his times, used
to manifest itself. It is not surprising, therefore, that he was
mainly looking for traces of science, law, market economy, and
other similar accepted attainments of civilization, and that in
view of the fact that the search had to be conducted under
different conditions it had to lead to broadened interpretations
of the various concepts.

There is one point more to stress. The dogma of the rationality
of the primitive man was, as mentioned above, linked to the
conception of culture as an instrument of survival. That conception naturally led to the question how primitive man could
have survived if, as was claimed by Levy-Bruhl, he had been
marked by a pre-logical (i.e., irrational – at least in the light
of current interpretations) mentality. For Malinowski – as a
result of the assumptions mentioned above – this question
amounted to asking: how could a primitive culture provide
conditions for survival without developing, at least in a nuclear
form, such powerful instruments of survival as those at the
disposal of civilized man, including precisely science? In this
way theoretical reasons made Malinowski adopt the purely
empirical thesis that science is a universal phenomenon.

It can thus be said that Malinowski's concept of science
is conditioned by his theoretical system and that it plays an
important role in that system. If only for this reason, his
conception is an essential element in social science in general,
and in the theory of culture in particular.

Florian Znaniecki wanted to pursue an 'inductive nonevaluative
science of knowledge',(11) i.e., wanted to view knowledge
'historically' and to 'compare the history of knowledge with
that of other domains of cultural achievement.'(12) He assumed,
'from experience and observation, direct and indirect', that
'knowledge as it has historically grown is the agglomerated
product of specific cultural activities of numberless human
individuals' and that 'some individuals for longer or shorter
periods of their lives specialize in cultivating knowledge, in
distinction from other individuals who specialize in performing
various other kinds of cultural activities – technical, economic,
artistic, and so on.'(13) Those individuals who specialize in
cultivating knowledge are 'men of knowledge' or 'scientists'.
But it soon turns out that this must be understood in a specific
way: a man of knowledge, or scientist, is the man who plays
the social role of a man of knowledge. This seemingly small difference in formulation is of essential significance here: one may

play the social role of a man of knowledge without being a man
of knowledge according to some 'objective', e.g., epistemological,
criteria. The social role of a man of knowledge is played by that
man who cultivates that which is treated as knowledge in a
given society. Hence for Znaniecki

> all knowledge is valid which is regarded as such by the
> people who participate in it, and a 'scientist' is any indi-
> vidual who is regarded by his social milieu and who regards
> himself as specializing in the cultivation of knowledge,
> irrespective of the positive or negative judgement which
> epistemologists or logicians pronounce upon his work. (14)

Thus the concept of knowledge is here by definition linked
to that of the social role of the scientist. Because in social
situations it is much easier to identify social roles than the
content of social consciousness, this linking as a rule makes
one adopt the social role as the starting point in conceptual
explanations, i.e., in following the rule: science is what is
being done by scientists. This is also what Znaniecki did in
practice: the problems of knowledge, and hence those of
science, which will be commented on below, became for him a
branch of the theory of social roles. True, Znaniecki did accept
the existence of 'knowledge without scientists', but he concerned himself with the knowledge of scientists by analyzing
the various, historically changing, social roles of scientists:
that of the sage, that of the scholar and that of the explorer.
The various forms of knowledge were thus correlated with
specific social roles.
Historically, the latest form of knowledge is connected with
the role of the scientist in a narrower sense of the term. It
is not clear whether Znaniecki's intention was to term that form
of knowledge, science. Some passages in his work speak in
favour of such a convention, others, in favor of linking the
concept of science to that of scholar in general, still others,
in favor of identifying science and knowledge (Znaniecki often
used these terms interchangeably). This, however, is not very
important: whatever interpretation we adopt, science remains
a domain of culture treated in terms of the theory of social
roles. (15)

It may be that the above examples do not testify to the theoretical refinement of the social conceptions of science, but they
show that such conceptions exist and, in particular, that they
are often an integral element of theoretical systems (however
we might assess such systems from a substantive or methodological point of view).
It is easy to prove that Malinowski's conception like his
whole theoretical system, suffers from naivety and is partly
based on wrong assumptions, (16) as well as that Znaniecki
failed to remain within the sphere of social issues and from

time to time resorted to highly debatable epistemological arguments. Such conclusions, even if they could be generalized (and
I think to a large extent they can), should not make us replace
imperfect social conceptions by philosophical conceptions, even
perfect ones. Contrary to Lakatos's belief, a normative philosophy of science as answer to the question about the essence
and limits of science, cannot change sociological and historical
discussions of science from what he calls 'confused rambling'
into discussions at a higher level, because the concept of
science defined in this way would be rather unnatural from the
point of view of social issues, and hence would doom that set of
problems to sterility. If we want to discuss science in the context of social problems, then the concept of science must
emerge from those problems: it must be a product of thinking
in social and cultural terms, and not in terms of logical connections among sentences. Lakatos says that the discussion of
whether science is a purely European phenomenon, is valueless
as long as science is not defined in terms of a normative philosophy of science. It seems that the contrary holds true: such
a discussion can have significant value only as long as science
is not defined in terms of any normative philosophy of science.
The problem of the historical and cultural limits of science
understood in accordance with the assumptions of inductivism
or Lakatos's methodology of scientific research programs may be
an interesting curiosity, especially for those who adopt the
appropriate methodological model, but it is not clear what such
a problem could mean for a student of social facts. The belief
that science, defined in accordance with the assumptions of
some normative methodology or other, is a correlate of actual
social facts requires proof, which – one can say in advance –
is not an easy undertaking as definitions constructed in that
way are notoriously inadequate not only in relation to the social
history of science but also in relation to research practice,
which has been confirmed by the fortunes, or rather misfortunes,
of the various normative methodological systems. Of course, such
inadequacy could not be objected to if only normative systems
in the strict sense of the term were concerned, i.e., methodological systems which do not aim at describing actual research
practices. But then it would be pointless to ask about the
historical and cultural limits of science.

The crux of the matter, however, is that every definition of
science, whether philosophical or other, is an artifact, which
in this case does not mean a negative evaluation.
    According to a certain tradition, connected mainly but not
exclusively with the history of logical positivism, scientific
operations are unthinkable without appropriate definitional measures. Schlick wrote that 'any inquiry must be preceded by
some kind of definition of the area that is to be studied',(17)
thus formulating one of those methodological assumptions of
logical positivism which have won the status of an obvious truth

in much wider circles. The interpretation of this assumption
used to change alone with the complications of the conception
of definition. More precisely, the contents of the interpretation
were becoming increasingly liberal as the rules for explaining
the meaning of scientific terms were being liberalized (the path
led from a direct explication by means of a full definition to
complex explication procedures which indirectly linked the term
being defined to the theoretical system in question). This
resulted in a situation in which Schlick's formulation is inaccept-
able: first, because what is required is not so much to define
the terms used as to make such terms bear specific relations to
a given system of language; (18) second, because in science
such relations only in limiting (and rather uninteresting) cases
can be formulated in a comparatively simple manner, namely as
a complete definition or a system of partial definitions, and
usually are rendered as is believed now, by the set of assumptions of the theoretical system in which they occur.

Liberalization in that field did not, and could not, go so far
as to accept, without additional explicative operations, such
notoriously vague terms as 'false consciousness', 'the superego',
'charismatic authority', not to mention something as scandalous
as 'Zeitgeist'. These terms, it is true, occur in definite theoretical systems marked by definite assumptions, but those
assumptions do not form a set of sentences which could be
assigned, without obviously straining the facts, an explicative
function relative to those terms. In other words, these terms
are defined not by sets of assumptions of the respective theoretical systems, but by those systems as wholes, with the proviso that the structure of such systems is evidently less clear
than the structure of those systems which, in the philosophy
of science, are identified with scientific theories. Hence the
interpretation of said concepts required operations incompatible
with the criteria of intersubjective controllability of the process
of research as understood by logical positivists; the problem
can only be solved by appropriate explicative operations, which
at best means a modification of such concepts.

Concern for proper specification of the meanings of the terms
used is due to the justified conviction that one has to know what
one is talking about, and that this knowledge should be objectified if we are to communicate with others. But what is not
noticed is that such objectification can take on the most varied
forms: a simple formal definition, a complex intellectual system,
or societally transmitted behavioural patterns. Special preference
for one of those forms, e.g., definitions, can be justified by
technical considerations (e.g., precision and efficiency), but
is admissible only within the limits marked by substantive considerations. Accepting only simple forms of objectification of
terminological conventions must result in either a drastic narrowing down of interests or in producing artificial constructs.

Logical positivism mainly brought about consequences of the
latter kind, at least in the domain of the social sciences. The

process of 'improving' the conceptual apparatus of the social science (which was supposed to give terms precision and to remove their semantic penumbra) resulted in the production of artificial conceptual entities, i.e., concepts which lacked a proper substantive background. Transforming the conceptual apparatus of complex intellectual systems so as to meet the requirements of an explication assumed to be simple and precise, yielded a network of concepts which was in fact founded not so much on the introductory systems (which were as a rule substantively poor) as on the old conceptual apparatus, i.e., on those intellectual systems which violated the requirements of technical tidiness. Thus, for instance, the term 'social class' has innumerable more or less precise and coherent definitions (many of them adjusted to the needs of survey studies), but its permanent status in sociology is not due to those definitions, but to the semantic penumbra it has acquired by operating in classical sociology, and especially in Marx's system. If we accept the various definitional explications of the term 'social class' and if it seems to us that they introduce important categories, then this is because these categories bring their prototype to our minds. If we remove classical sociology from its setting and treat those explications literally, we are left with a set of arbitrary conceptual constructions and a mass of loose items of information which turn out to be astonishingly insignificant.

Matters become still more complicated when we consider such concepts (with their corresponding terms) as the Renaissance, humanism and Western civilization, which are not linked to any definite theoretical system (although their corresponding terms may, of course, have specific interpretations in the various systems), and whose respective meanings are determined by the whole of our social and historical knowledge and change as that knowledge changes. The difference is essential. The concepts determined by a theoretical system are more or less closed, at least to a large extent, and their changeability, like that of the entire system, is limited by the basic assumptions of that system.(19) The term 'Renaissance' is an open one: the Renaissance is a name of something which in the light of definite social and historical knowledge emerges as a certain historical whole; what belongs to that whole, what its fundamental features are, and even what its spatio-temporal coordinates are, is not specified by any definition nor any theoretical system. In other words, it is a completely open question. The concept of the Renaissance thus logically precedes its theoretical definitions: the answer to the question, what is the Renaissance, is inherent in the objective historical consciousness of a given epoch, and every serious theory of the Renaissance takes that answer as its point of departure, even if it contributes to modifying it.

It is more or less the same with the concept of science. A concept of science exists, independent of any particular theory

of science. That concept is determined by the context of all
our actual social and historical knowledge and results from the
structuring of the world of culture, imposed by that knowledge.
Thus, that concept has the same objective existence as the
concepts introduced definitionally or by assumptions of a theory,
but in an incomparably more complex form, which is therefore
much more difficult to interpret. Therefore, comprehension of
what science is, is a matter of the appropriate humanistic
competence and strictly depends on a comprehension of culture
in general. As a result, that concept is better or worse understood, in the same way as people better or worse understand
what religion, magic, art, etc. is. Generally speaking, comprehension (or grasping) of the concept of science requires comprehension of the context which determines it.

That context changes, of course, and so accordingly does
the concept of science. Discussions of the essence, limits, etc.,
of science are empirical in nature: their point is not how
'science' is to be understood, but how to grasp certain fundamental properties of a certain emerging whole and of its place
in the structure of cultural facts. The question 'what is
science?' is a question about the structuring of the world of
culture, and splitting it into a question about the meaning of
the term and a question about the properties of its designatum
would be both technically absurd and theoretically wrong. To
return to the example which scandalized Lakatos, a historian
who reflects on whether science is a specifically European
phenomenon, in equal measure shapes the concept of science
and establishes historical truth. The research carried out by
Needham's school modified both the concept of science and
our knowledge of the culture of the Far East, and separation
of these two aspects is impossible even in principle.

Thus the concept of science we have been speaking about is
indefinable in the very strong sense of the word, i.e., it is
essentially indefinable. This is because it is an open concept,
i.e., one which is not limited by any system of theoretical
assumptions and such that its shape is determined by the changing picture of the world of culture.

Let us note, however, that in reflections on the history or
anthropology of science, the arguments in favour of the scientific nature of an intellectual undertaking usually tend to demonstrate its rationality. Does this not mean that in spite of everything people act in accordance with Lakatos's idea of a good
historian of science, i.e., adopt a normative methodological
system which determines the criteria of rationality of cognitive
operations and thus makes it possible to distinguish scientific
operations from nonscientific ones?

Discussions of the rationality of some intellectual enterprise
of course assume a system of criteria of rationality, but this
has no significance for the role of the philosophy of science,
or any other conceptions of rationality in such discussions. In

particular cases we may, of course, use criteria of rationality
advanced by some philosophical doctrine or other, but on the
whole we use those criteria of rationality which are imposed
upon us by our culture. What is more, the criteria of rationality
imposed upon us by our culture are primary in relation to the
philosophical conceptions of rationality. Every such conception,
regardless of where it seeks support, is merely an attempt to
articulate the criteria which are actually valid in a given culture,
and only this gives such a conception its real support. Popper
is of course right when he claims that no doctrine of rationality
can be substantiated; this is not to say, however, that every
such doctrine operates in an intellectual vacuum. The general
system of ideas which combine to form our culture is always the
final arbiter. Both the anthropologist who reflects on the rationality of magical belief and the philosopher who constructs a
methodological doctrine, refer to that system whether they are
aware of the fact or not. For both of them it is the ultimate
source of criteria of rationality and both avail themselves of it
in the same way: by their feel for the norms involved in the
context of the whole culture. The difference between them is
that the anthropologist transforms the feel he has into an
interpretation of a given case, while the philosopher transforms
it into a more or less orderly doctrine. The secondary nature
of such a doctrine relative to the norms involved in the whole
of culture becomes especially apparent at times when the doctrine experiences a crisis: this is always a period of evident
and no longer defensible discrepancy between the doctrine and
the broader intellectual context of the epoch. I think logical
positivism has been in this situation for over a dozen years.
Belief in the intellectual autonomy of philosophical doctrines
which propagate some such system of criteria of rationality,
treating them as independent and necessary instruments for
evaluating intellectual activity, results from a lack of the
proper historical perspective when viewing the philosophical
analysis of knowledge. Feyerabend's slogan that 'anything
goes', in so far as it rejects the whole normative methodology
of science, does not, as Popper feared, open the door to arbitrariness; rejection of all the doctrines which stress rules of
rationality does not amount to a rejection of all rules of rationality. For the same reason, the objection raised again Kuhn,
in particular his 'The Structure of Scientific Revolution', that
he advocates 'mob rule' in science, is a misunderstanding: the
behaviour of that mob may be a result of exemplary subordina-
tion to certain rules - which, however, are changing and not
visible from the perspective of the philosophy of science.

Thus the problem of rationality and their connections with
the concept of science do not force one to modify the point
about the openness of that concept. Science is, in fact, always
required to be a rational undertaking; it is that requirement
which defines the boundaries that separate it from other spheres
of culture, but as long as the criteria of rationality are deter-

mined by culture, and not by some conception of rationality, the concept of science remains indefinable.

From this point of view, every definition of science, whether a simple formula, a methodological doctrine, or a sociological theory, is a construct that can be more or less useful but does not grasp the essence of science. Of course, it may always be stated that one only has the introduction of a certain termino-logical convention in mind, but first, such a statement would be false regardless of its author's intentions, and second, one would have to prove the substantive value of such convention, which is never done (just because people practically always mean 'science' in its primary, culture-based sense, and hence a concept whose raison d'être is self-evident).

The philosophy of science could be 'the history of social methodo-logical consciousness'. This expression of Jerzy Kmita's, (20) I interpret in accordance with what has been said earlier: the philosophy of science could become a part of the social and humanistic studies of science, i.e., studies in which science is interpreted in the context of a general reflection on culture. Philosophy of science cultivated in this way would have to be a historical discipline, i.e., a discipline whose cognitive perspec-tive excludes belittling the changeability of the subject matter of study, changeability observable in the history of culture. It would also have to have at its disposal methodological pos-sibilities that not only go far beyond the needs of 'logical analysis', but also the standard equipment of the present-day 'empirical sociology of science'.

Philosophy of science conceived in this way exists only in the form of more or less consciously formulated requirements, sup-ported, on the one hand, by the crisis in the present-day philosophy of science, and on the other, by some publications, especially those by Kuhn and Feyerabend, and in Poland by Amsterdamski, Rainko, and the Poznań School. But the main trend in the philosophy of science has not been modified since the times of the Vienna Circle and is a quite different under-taking, intended to yield a theoretical construction which could be accepted as an adequate methodological model of science. This model is to be universal by assumption, i.e., one which shows the methodological properties of science in general, and not just some of its forms (e.g., contemporary physics). In this sense the philosophy of science is pronouncing about the future of science: it is logically impossible that science could in the future be fundamentally different in its methodology from con-temporary science, for such differences would mean that science had changed into an undertaking of a different kind.

Theoretically speaking, this goal of an adequate methodological model could be attained in various ways: some of them might quickly result in the abandonment of that undertaking as unrea-listic, while others might preclude such a possibility in advance. The way adopted by the Vienna Circle and used so far in the

philosophy of science comes close to the latter extreme, which
means that it produces strong safeguards against any need to
revise the goal of action.

The logical positivist philosophy of science was by assumption
a theoretical description of a section of *reality* termed 'science'.
This 'naturalistic', as Popper used to call it, conception of the
philosophy (methodology, theory - these terms have been used
alternately by logical positivists) of science has probably been
most clearly formulated by Reichenbach:

> Every theory of knowledge must start from knowledge as a
> given sociological fact. The system of knowledge as it has
> been built up by generations of thinkers, the methods of
> acquiring knowledge used in former times or used in our
> day, the aims of knowledge as they are expressed by the
> procedure of scientific inquiry, the language in which
> knowledge is expressed - all are given to us in the same
> way as any other sociological fact, such as social customs
> or religious habits or political institutions. The basis avail-
> able for the philosopher does not differ from the basis of
> the sociologist or psychologist; this follows from the fact
> that, if knowledge were not incorporated in books and
> speeches and human actions, we never would know it.
> Knowledge, therefore, is a very concrete thing; and the
> examination into its properties means studying the features
> of a sociological phenomenon.(21)

A well-expressed and pertinent formulation. But two pages
later we read that:

> Epistemology does not regard the processes of thinking in
> their actual occurrence; this task is entirely left to psycho-
> logy. What epistemology intends is to construct thinking
> processes in a way in which they *ought to occur if they are
> to be ranged in a consistent system*; or to construct justifi-
> able sets of operations which can be intercalated between
> the starting-point and the issue of the thought-processes,
> replacing the real intermediate links. Epistemology thus
> considers a *logical substitute rather than real processes*.
> For this *logical substitute* the term *rational reconstruction*
> has been introduced ... *It will therefore, never be a per-
> missible objection to an epistemological construction that
> actual thinking does* not conform to it.(22)

An about-face? Yes, and no.

> In spite of its being performed on a fictive construction,
> we must retain the notion of the descriptive task of epistemo-
> logy. The construction to be given is not arbitrary; it is
> bound to actual thinking by the postulate of correspondence.
> *It is even, in a certain sense, a better way of thinking than
> actual* thinking.(23)

This astonishing dialectic is necessary in this case. The conception of the philosophy of science advanced by logical positivists had to bring conflicting trends into harmony. The point was, one, to formulate a theory of science that would correspond to reality; two, to do so by means of logical analysis; three, to provide a theory having the properties of a normative system with the already known consequences (the superiority of science over the other domains of intellectual activity, the superiority of the natural sciences over the social sciences, the relative perfection of physics, etc.). Carnap's conception of rational reconstruction was in fact the only way of avoiding a situation in which one would have to choose among the various assumptions, e.g., between research techniques and descriptive aims. This is why it was accepted universally and exceptionally uncritically (if we discount outsiders' opinions, which could not significantly influence the insider's discussions, especially since such opinions usually questioned the value of the entire undertaking, and not just the various assumptions). The differences of opinion were reduced to differences in acceptance of the various points: for some, rational reconstruction was chiefly reconstruction, for others (mainly for Popper) it was above all rational. To this day, the philosophy of science is practised in accordance with these same assumptions despite inner criticism directed at making the philosophy of science history-oriented (Kuhn) or psychology-oriented (Quine).(24) This is understandable, insofar as rejection of Carnap's idea of 'rational reconstruction' would either result in a complete methodological reorientation (with a subsequent change of research methods) of the philosophy of science or in opening the question of the subject matter of that discipline, i.e., in asking whether that is the philosophy of science, and if not, whether it is the philosophy of something interesting.

The fundamental problem of the philosophy of science in the form it has had so far is the contradiction between research methods and descriptive aims. Its methods include a logical analysis of what science produces (i.e., statements made in science). Note that the wording is ambiguous. In a sense, a logical analysis of what science produces is the business of every researcher, who must continually evaluate, criticize, etc., the correctness of reasoning, the validity of conclusions, the coherence of conceptions, etc., who, in a word, must think logically. But the logical analysis used in the philosophy of science is logical analysis in a stronger sense, namely an analysis in terms of logic (as a discipline). We are thus dealing with an approach to science permitted by the apparatus of contemporary logic. This apparatus is obviously stronger than its classical predecessor - a point so forcefully stressed by Carnap - and, more important still, it is being expanded, among other things, in connection with the needs of the philosophy of science. But, in the first place, those needs are determined more by the inner problems of methodological doctrines than by

what takes place in the laboratories of physicists and biologists
(even if we assume that philosophers are aware of what is taking place there), and in the second place, there is - in a very
practical sense - the problem of whether the contemporary
system of logic is adequate to describe science, especially if
one does not assume in advance that scientific production is
methodologically homogeneous.

Adequacy of the tool is one thing, and how one uses it is
another thing - and more important in this case. The essence
of 'rational reconstruction', as usually applied, is determined
by the point of departure. It is usually a trivial logical problem
which is intuitively believed to be an important issue in the
logical structure of science and which is accordingly analyzed
in the appropriate linguistic version (explanation, prediction,
theory structure) and illustrated by examples drawn either
from every-day experience (the freezing of water in the cooler)
or from textbook physics. More complicated examples came to
be analyzed only recently, under the impact of the 'history-
oriented' trend. Furthermore, such a problem is analyzed from
the point of view of certain epistemological assumptions in the
traditional sense of the term. Thus, for instance, the beginning
of logical-positivist philosophy of science was dominated by the
problems of the logical connections between observation statements (Protokollsätze) and other statements made in science.
The two fundamental questions were: what falls under the category
of observation statements and what are the connections between
them and other statements made in science. But neither question
had an empirical interpretation nor required verification as to
how things really stand in science. The former was an epistemological question and really meant: how is the concept of observation statement to be interpreted from the point of view of the
epistemological assumptions of empiricism;(25) the meaning of
the latter amounted to the question: what logical connections
must there be between observation statements and other statements, if the acceptance of the latter is to be justified by the
acceptance of the former?

The connection between these problems and the real problems
of the logical properties of scientific systems was merely a
matter of making a statement, which Reichenbach euphemistically
called 'the postulate of correspondence'. If so, then we have
to agree with Feyerabend that in the philosophy of science
'"Problems of science" ... are the internal problems of the
chosen (logical) system, or set of systems, illustrated with the
help of bowdlerised examples from science itself.'(26) With two
reservations, however. First, the formulation of these problems
was essentially influenced by the assumed epistemological doctrine, especially in the initial period in the development of the
philosophy of science.(27) Second, 'the postulate of correspondence' though not binding in the short run, started to operate
when a discrepancy between some model of the philosophy of
science and the facts was becoming evident. The logical-

positivist philosophy of science could not, for doctrinal reasons, be a theory of rational knowledge; scientific knowledge was by assumptions the only form of rational knowledge, and hence one's own ideas of the rules of rational knowledge had to be brought into harmony with one's own ideas of the properties of scientific knowledge. In my opinion, this explains the unprecedented evolution of methodological doctrines during the last fifty years. It is common knowledge that this evolution consisted in a liberalization of requirements. Note also that it was a process during which the picture of the methodological properties of science was becoming more and more complicated. One of the sources of that complication is certainly to be seen in the inner problems of logical systems (e.g., the paradoxes of probabilistic explanation), another one – more important in the long run – in the problems which imposed themselves during even the most cursory observation of what is really going on in science (e.g., the problem of the symmetry of prediction and explanation). It is self-evident that the change which has taken place in the logical picture of science during the last fifty years would have been impossible if the said 'postulate of cor-respondence' did not force researchers to go beyond the inner problems of methodological models.

It is a different question, to what extent that process can be treated as an approach to the real picture of science. Does Lakatos's methodology of research programmes really come closer to what is really going on in science than Carnap's methodology did in the 1920s? Feyerabend's answer to the question, is in the negative. In the general sense, it is 'no', because science does not observe any methodological rule, and cannot do so if it is not to destroy itself. The only universally observed rule is that anything goes, so that any methodological model is as much a caricature of science as one of its rivals.(28) To cite a particular case:

> The methodology of research programmes most certainly has led to some interesting historical discoveries. This is not surprising. Any hypothesis, however implausible, can widen our horizon. It has not led to a better understanding of science and it is even a hindrance to such a better under-standing because of its habit of beclouding facts with sermons and moralizing phrases.(29)

Like every true philosopher of science Feyerabend, when he says 'science' is of course mainly thinking of present-day physics, and considers the issue only in that context. Still, what he did is an important fact whose significance is, I think, this: following a circuitous path, as it were, the philosophy of science has reached the point at which the question of whether the methodological picture of science agrees with the facts is recognized as a problem. Hence this is the point at which the conception of the rational reconstruction of science has in a

practical sense proved to be the weakest. A breach made here
may bring about a basic revision of the methodological foundations of the discipline in question, and in particular may result
in incorporating empirical studies as an indispensable element
of the entire undertaking. Such a reorientation would automatically change the status of the great methodological systems of
our times by reducing them to the level of theories which are
evidently false.(30) On the other hand, as the case of Kuhn and
other representatives of the history-oriented trend shows, such
a methodological breach does not, and cannot, automatically
change the foundations of that discipline, and in particular the
system of assumptions on which the philosophy of science came
to be based at the time of the Vienna Circle. I chiefly refer here
to:

(i) posing a conflict between science and the other spheres
of intellectual activity in the belief that this follows a (clear)
demarcation line between the sphere of rational knowledge and
one (possibly) representing other values;
(ii) the belief that science as a whole is marked by historically
constant methodological properties that form its non-trivial
characteristics;
(iii)   the belief that all major methodological differences
between the various disciplines can be explained as differences
in the degree of development;
(iv) treating physics as the most advanced and thereby
model discipline.

The methodological reorientation which would consist in
incorporating serious empirical research on the methodological
foundations of present-day physics would not, of course, change
this system of assumptions, and thus would not change the fact
that the gap between the philosophy of science and the contemporary sciences of culture is growing wider and wider. The
philosophy of science was based on a system of beliefs concerning the origin and nature of certain spheres of culture, beliefs
that were current in the early twentieth century. Those beliefs
are increasingly difficult to defend, if not outright anachronistic
This is why, I think, the philosopher of science can derive
more advantage, e.g., from a discussion of the rationality of
magic than from empirical studies on the real methodological
properties of physics during the last two hundred years. Generally speaking, what would most affect the further development
of the philosophy of science would not be making it history-
oriented in the sense advocated by Kuhn, i.e., confronting it
with certain empirical data (very limited in scope), but confronting it with the rival theoretical interpretations of science
to be found in present-day theoretical reflection on culture.
In other words, the philosophy of science not only needs a
new methodological orientation, but also, and primarily, a new
theoretical orientation, since otherwise, regardless of the

relation it would bear to the research practice of the contemporary physics, it would remain burdened with the old dogmas
that stand in the way of a sound discussion of such issues as
the methodological changeability and differentiation of science,
or the cultural and social conditionings of the criteria of
rationality. If it is assumed in advance that there is only one
changeless model of rational knowledge and that it can most
easily be found in present-day physics, then regardless of
the research methods employed one is unable to notice anything
taking place outside that domain. In the next chapter we shall
see how this mechanism of selective perception works in
practice.

4 The unity of science as a 'logical necessity'

Otto Neurath, in his 'Empirische Soziologie' wrote (underlining that sentence) that 'All genuine science can only be physics.' He was also convinced that 'there is no question ... that there is only one kind of physics, the basis of which is controlled by "observation statements".'(1) This, together with the assumptions concerning observation statements and their subject matter, easily led to the conclusion that 'What appears in statements as "mental", "personality", "social", must be expressible as something spatio-temporal or else vanish from science.'(2) As we know, in the Vienna Circle terminology to vanish from science meant to remain outside the sphere of rational knowledge, to be the subject matter - should anyone so wish - of emotion, but not of thought.

Argumentation constructed more or less in this form was a typical element of the 'philosophical analysis of the language of science' when it came to the language of the social sciences. While the attitude of the Vienna Circle toward the natural sciences, and especially toward physics, was based on the assumption that they embodied the essence of science which made its members bring their methodological models into agreement (however this was to be done) with the findings of the natural sciences, the attitude toward the social sciences was based on the assumption that they were basically pseudo-scientific disciplines in need of radical methodological reforms.(3) Thus, while the scientific status of physics was predetermined, the scientific character of the social sciences remained an open issue, though in practice there was no doubt that a 'logical reconstruction of science' could find only negative examples there.

Note that this asymmetry is not substantiated by the system of basic assumptions of the logical-positivist philosophy of science. On the contrary, it is at variance with the conception of the philosophy of science as a discipline which investigates the methodological properties of science as a sociologically given fact, because that sociologically given fact covered both groups of academic disciplines equally, especially for the members of the Vienna Circle, who mostly used the German language and thus had no linguistic reasons to identify science with natural science. Nor was this asymmetry adequately supported by the epistemological assumptions of empiricism. Neither those assumptions nor interpretation of them adopted by the Vienna Circle provided grounds for such an unequal treatment of the

various academic disciplines. Immediate empirical data, observation statements, observation terms, etc., could be sought as
well in physics as in sociology, and with the same effect. In
more general terms, the unequal treatment of natural science,
and especially physics, on the one hand, and the social sciences,
on the other, was introduced into the philosophy of science by
chance as a social consequence of generally held ideas and
valuations, and not on any substantive grounds. But the chance
factor, once introduced, rapidly produced rationalizations to
support it: 'the logical reconstruction of science', restricted by
the assumptions adopted in advance, obviously demonstrated
the correctness of such evaluations. If one had such great
opportunities for not taking facts into account as were provided
by the idea of rational reconstruction, and if one firmly believed
that metaphysics is a result of intellectual aberration, that
sociology is a pseudo-science, and that physics is the paragon
of rational knowledge, one could hardly arrive at conclusions
that would undermine that conviction. Here we come back to the
point at which the philosophy of science must be viewed as a
sui generis rationalization of every-day experience. That experience makes one rank physics higher than biology, and biology
higher than sociology. This ordering is certainly informative
about the social prestige of those disciplines, but there is no
reason to treat that ordering as if it were based on some reason-
ably pertinent, even if intuitive, comparison of the substantive
values of the various disciplines. It is another matter whether
such a comparison is even possible: this issue will be discussed
later, and for the time being we confine ourselves to stating
that one of the représentations collectives was uncritically
introduced into the philosophy of science as a self-evident
truth.

Be that as it may, the logical-positivist philosophy of science
did not make any serious effort to investigate the methodological
properties of the social sciences, which is not to say that
logical positivists did not pay any attention to them. In that
respect it is worth singling out two periods: the first covers
the 1920s, the 1930s and the early 1940s, and the second the
last thirty years. The first was marked by an exceptionally
sharp, one-sided and schematic criticism of the situation prevailing in the social sciences and of traditional (especially German) philosophical reflection on the foundations of those
sciences. The second was marked by a lack of interest in the
social sciences (the attention of philosophy was focused almost
exclusively on physics), with the declarative support of the
earlier general theses on the issue. The first period is more
interesting for us, not only because the theses on the social
sciences which have retained their validity in the philosophy of
sciences were formulated at that time, but also and primarily,
because that period saw the strongest impact of the philosophy
of science upon sociology and related disciplines. The opinions
which were adopted then, very soon acquired the status of

methodological foundations of the social sciences and came to function somewhat independently of the further evolution of the philosophy of science.

In his paper on the logical analysis of psychology, first published in the Revue de Synthèse in 1935, Hempel suggested, as he claimed, a new trend in the analysis of psychology, an approach which 'makes use of rigorous logical tools' and offers opportunities for resolving the well-known problem: whether psychology is a natural science or one of the Geisteswissenschaften. The new approach consists in the logical analysis of the language of science, 'which became possible only with the development of an extremely subtle logical apparatus which makes use, in particular, of all the formal procedures of modern logistics.'(4) Let us see how Hempel analyzed 'the language of psychology'. His argument followed the pattern then being applied by the Vienna Circle. He began by presenting a certain - current, as he claimed - conception of the humanities, which implies the thesis 'that there is a fundamental difference ... between the natural science as a whole, and the sciences of mind and culture',(5) and later used that conception as polemic from which background he presented the standpoint of the Vienna Circle. That conception, he said, takes on diverse formulations, but behind that diversity there is one common idea which can be described thus:

Apart from certain aspects clearly related to physiology, psychology is radically different, both as to subject-matter and as to method, from physics in the broad sense of the term. In particular, it is impossible to deal adequately with the subject-matter of psychology by means of physical methods. The subject-matter of physics includes such concepts as mass, wave length, temperature, field intensity, etc. In developing these, physics employs its distinctive method which makes a combined use of description and causal explanation. Psychology, on the other hand, has for its subject-matter notions which are, in a broad sense, mental. They are toto genere different from the concepts of physics, and the appropriate method for dealing with them scientifically is that of sympathetic insight, called 'introspection', a method which is peculiar to psychology.(6)

Who held such opinions? Now Hempel did not give any name or bibliographical data, but there is no doubt that he meant the antipositivistic opposition in the philosophy of the social sciences and humanities, especially the various theories of Geisteswissenschaften. That opposition, while greatly differentiated, was linked together by a certain common system of fundamental assumptions, but anyone who knows that trend, be it only from cursory reading, also knows that it is not so unattractive and shallow as the explanation given above would lead us to believe.

This is extremely characteristic: one of the most striking features of logical-positivism, especially, though not only, in its
initial period, was an extraordinary, when judged by academic
standards, dishonesty with respect to the opinions of opponents,
which in particular consisted in reducing those opinions to
trivial and artificial constructions. This applies primarily to
philosophical reflection on the foundations of the Geisteswissenschaften. Thus, for instance, the whole problem of 'Verstehen'
was stereotypically presented as an empathy in other people's
experiences, which led to the conclusion that Verstehen played
a merely heuristic role. As Otto Neurath wrote, '(it) enter(s)
the totality of scientific statements as little as does a good cup
of coffee which also furthers a scholar in his work.'(7) Ana-
logously, the interpretation of culture as geistige Welt was
transformed into a conception which reduces the problems
investigated by the social sciences and humanities to mental
experiences. Historicism, as interpreted by Popper, not only
has no interesting connections with studies by Friedrich Mein-
ecke or Dilthey but as Robert Ackerman has noted, it is 'an
unattractive set of ideas that probably no one has ever held.'(8)

These kinds of interpretative operations have, unfortunately,
resulted in a complete vulgarization of the achievements of
antipositivist reflection on the humanities and, in view of the
popularity of logical positivism, in a practical elimination of
that tradition from the social sciences, or at least from sociology and its related disciplines. Papers on, and handbooks of
methodology circulated conceptions which resemble the antipositivist traditions only by name, and yet have effectively
managed to replace that tradition in the consciousness of sociologists.

But let us return to what Hempel wrote. The Vienna Circle,
he said, rejected the opinions formulated above and stood for
the unity, both as to the subject matter and methodology, of
all scientific disciplines. The substantiation of that standpoint
was as follows: the theoretical content of science is to be found
in propositions.(9) Therefore we have to find out whether there
is an essential difference between psychological statements, on
the one hand, and statements in physics, on the other. Now
there is no such difference because:

(i) 'the meaning of a proposition is established by the conditions of its verification,' which reduces the problem to the
question of the conditions of verification of statements in
psychology and in physics;
(ii) these conditions are identical: in both cases they take
on the form of protocol statements.(10)

Now (i) expresses the then held standpoint of the Vienna
Circle concerning 'theoretical statements'. Each such statement
was treated as reducible to a set of protocol statements:

A proposition that specifies the temperature at a selected point in space-time can be 'retranslated' without change of meaning into another proposition - doubtlessly longer - in which the word 'temperature' no longer appears. This term functions solely as an abbreviation, making possible the concise and complete description of a state of affairs, the expresion of which would otherwise be very complicated.(11)

Hempel of course assumes that theoretical statements in psychology are also reducible to protocol statements, and moreover, which is the key issue in this context, that protocol statements are in this case also physical statements. He accordingly formulated the general characteristic of psychological statements (his original text being in italics):

All psychological statements which are meaningful, that is to say, which are in principle verifiable, are translatable into propositions which do not involve psychological concepts, but only the concepts of physics. The propositions of psychology are consequently physicalistic propositions. Psychology is an integral part of physics.(12)

Note that this formulation includes a significant restriction of the scope of the asserted thesis: reference is made not to psychological statements in general, but to psychological statements 'which are meaningful, that is to say, which are in principle verifiable.' It can easily be assumed that Hempel did not give any independent criterion that would make it possible to distinguish those statements which are meaningful and in principle verifiable from those which are not. In other words, from what has been said earlier it follows that this passage is a veiled definition of the term 'meaningful statement', which is in turn definitionally connected with the term 'verifiable statement'. Hence the truth of the claim included in that passage cannot be disrupted: a psychological statement which would prove not to be a physicalistic statement would thereby be meaningless, and as such would be outside the domain of (scientific) psychology.

This operation was repeated unchanged and with astonishing effectiveness from the very formation of the Vienna Circle: research procedures in the social sciences and those of their results which did not satisfy the methodological model included in the philosophical doctrine found themselves automatically outside the domain of science (possibly to return to it when the doctrine was liberalized). In this way, the truth of the doctrine in regard to the social sciences was invariably assured by definitional operations. Thus, all criticism of the model which cited facts missed the point: a discrepancy between the model of science and non-scientific procedures clearly testifies in favour of the model.

But it was a foregone conclusion that it was impossible for

all the social sciences to lie outside the domain of science. The Vienna Circle aimed at annihilating metaphysics (which Popper, as we know, did not do since he recognized the raison d'être of metaphysics as a distinct sphere of intellectual endeavour separated from science), but not the social sciences. Their criticism of the social sciences – and the extreme sharpness of that criticism – was necessitated by the evident discrepancy between the suggested model of science and the real state of things in the social sciences. But criticism could not turn into outright rejection of that sphere of intellectual endeavour, because that would be incompatible with current evaluation which, as I have repeatedly emphasized, were the starting point of the logical-positivist vision of intellectual output. That was why members of the Vienna Circle argued that:

(a) 'the social sciences are possible', which in this case meant that it was possible to cultivate them in accordance with the methodological doctrine of the Vienna Circle;
(b) this was, in fact, already being done, but unconsciously and in a manner which was, first of all unacceptable, and secondly gave the impression that something quite different was being done.

In other words, the point was to demonstrate that there is nevertheless a connection between the doctrine and the real state of the social sciences. Thus the correspondence postulate played a role here, too, but a quite different one from the role that it played in the case of natural science; there, an evident and intolerable discrepancy between the methodological model and the facts led to a revision of the model; here, the same situation made one undertake appropriate interpretative operations relative to the facts, operations intended to demonstrate that social scientists, contrary to all appearances, do talk in prose – at least those who deserve the name of scientists. Let us see how such argumentation is carried out by Hempel.

To show that psychological statements are translatable into physicalistic ones he analyzes the statement 'Paul has a toothache'. In his opinion this statement is translatable into a set of physicalistic statements which formulate the conditions of its verification. These include such statements as:

a. Paul weeps and makes gestures of such and such kinds.
b. At the question, 'What is the matter?', Paul utters the words 'I have a toothache.'
c. Closer examination reveals a decayed tooth with exposed pulp.
d. Paul's blood pressure, digestive processes, the speed of his reactions, shows such and such changes.
e. Such and such processes occur in Paul's central nervous system.

'This list', Hempel continues, 'could be expanded considerably, but it is already sufficient to bring out the fundamental
and essential point, namely, that all the circumstances which
verify this psychological proposition are expressed by physical
test sentences.'(13)

The proposition in question, which is about someone's 'pain',
is therefore, equally with that concerning the temperature,
simply an abbreviated expression of the fact that all its
test sentences are verified. ... It can be re-translated
without loss of content into a proposition which no longer
involves the term 'pain', but only physical concepts. Our
analysis has consequently established that a certain proposition belonging to psychology has the same content as
a proposition belonging to physics.

This is immediately followed by a generalization:

The above reasoning can be applied to any *psychological
proposition*, even to those which concern, as it is said,
'deeper psychological strata' than that of our example. Thus,
the assertion that Mr. Jones suffers from intense inferiority
feelings of such and such kinds can only be confirmed or
falsified by observing Mr. Jones's behaviour in various circumstances. To this behaviour belong all the bodily processes of Mr. Jones, and, in particular his gestures, the
flushing and paling of his skin, his utterances, his blood
pressure, the events that occur in his central nervous
system, etc.(14)

Note that Hempel's method of substantiating the thesis on
the translatability of psychological statements into physicalistic
ones, has enabled him to avoid contact with psychology. The
relationship between the statement 'Paul has a toothache' and
psychology is the same as that between the statement 'it is
raining' and physics. Yet this is the only statement Hempel
analyzed: this analysis is followed by the declaration that
every psychological statement can be translated into a set of
physicalistic ones in the same way. In other words, Hempel's
text is not supported by any serious analysis of any important
fragment of psychological knowledge, such as some significant
proposition drawn from a serious psychological theory. His
argumentation is, in fact, based on assumptions he had adopted
a priori, and 'the logical analysis of psychology' turns out to
be an explanation of the methodological doctrine adopted irre-
gardless of what is being done in psychology and illustrated by
a single pseudo-psychological statement. In other words, it is
an analysis of 'psychology' without any study of psychology.
Reality has the properties it is bound to have: 'if in psychology
only physicalistic statements are made, this is not a limitation
because *it is logically impossible to do otherwise*.'(15)

The thesis on the translatability of psychological statements into physicalistic ones was quickly abandoned, and there is no need to attack it, the more so as we are here primarily interested in the manner in which proclaimed opinions are justified. That manner has not changed in the philosophy of science to this day. The thesis on translatability was abandoned not as a result of a confrontation with facts, but because it turned out that it was not a logical necessity after all. To put it more precisely, it was replaced by the thesis on the reducibility of psychological statements to physicalistic ones, and this occurred when it turned out that the thesis on the translatability of 'theoretical' physicalistic statements into 'test statements' in physics was untenable and had to be liberalized.(16) This process was repeated with unmistakable regularity: the theses on the methodological properties of the social sciences were always a reformulation of the actually valid theses on the properties of physics. Cultivation of 'the logical analysis of the social sciences' was thus largely a mechanical job, a kind of translation of existing theses into a somewhat different language. This was why it did not enjoy much prestige and was pushed even further to the outskirts of the philosophy of science.(17)

Note also that Hempel did not confine his conclusions to psychology, but extended them, also declaratively, to sociology:

> The considerations we have advanced can be extended to the domain of sociology, taken in the broad sense as the science of historical, cultural and economic processes. In this way one arrives at the result that every sociological assertion which is meaningful, that is to say, in principle verifiable, 'has as its subject-matter nothing else than the states, processes and behavior of groups or individuals (human or animal), and their responses to one another and to their environment' [a quotation from Carnap] and consequently that every sociological statement is a physicalistic statement.(18)

This is further extended to all social sciences:

> *Every proposition of the above-mentioned disciplines, and, in general of experimental science as a whole, which is not merely a meaningless sequence of words, is translatable, without change of content, into a proposition in which appear only physicalistic propositions.(19)*

The concept of the unity of science has at least three versions: (1) that of the unity of the language of science (2) that of the unity of the theory of science, and (3) that of the methodological unity of science. Originally, each version identified science with physics, and thus was an appropriate modification of physicalism. Physicalism, however, soon proved to be both untenable and superfluous, and was driven out by a broader conception. This did not, of course, mean the end of the cult of physics: in the new context disciplines other than physics acquired their raison d'être (they ceased to be accidental products of a reckless division of academic labour), but physics retained its special place as - henceforth - the model discipline.

The concept of the unity of the theory of science, or, in Feigl's terminology, radical physicalism, did not play any essential role in the development of the philosophy of science, nor did it in any way influence the development of the social sciences. It was a sui generis revival of mechanistic materialism. It treated physics as a basic body of knowledge to which all (real, true, etc.) knowledge could be reduced, or from which it could be deduced. That conception, from the outset highly controversial in the philosophy of science itself, has never been given a precise formulation. Herbert Feigl, who consistently defended it, though with some reservations,(20) thought that it could be expressed as the thesis: 'The set of physical laws which enables us to deduce the facts of chemistry will also be sufficient for biology and psychology.'(21) This thesis, however, is capable of various interpretations. Generally speaking, the concept of the unity of science in that (radical?) version has never, even approximately been worked out in such detail as has the concept of the unity of the language of science.

At first, even at the time when Hempel wrote his paper, there was no reason to make a distinction between these two concepts. The belief in the complete definability of the concepts of (scientific) psychology and other non-physicalistic disciplines in physicalistic terms entitled one to believe in the complete translatability of non-physicalistic statements into physicalistic ones. In the light of the atomistic conception of the theory this belief settled the issue of the theoretical unity of science in favour of the thesis which we also find in Hempel's above discussed paper: 'All the branches of science are in principle of one and the same nature, they are branches of the unitary science, physics.'(22) This also eliminated the problem of the methodological unity of science. The differentiation of the sciences, i.e., the existence of the various disciplines, was a social problem (non-rational division of labour) and in part also a technical one (as would follow from Hempel's argumentation, psychology would be a particular case of the aggregation of information on physics, and hence the issue was that of data aggregation).

Matters became complicated when it turned out that at least some physicalistic terms, namely those which Carnap labelled dispositional terms, are not definable in terms that refer to immediate empirical data. Hence the thesis that all extralogical scientific terms can be defined in terms that refer to immediate empirical data was replaced by the thesis that all extra-logical scientific terms are reducible to terms that refer to immediate empirical data, that is, to a weaker thesis, since reduction

meant explication either by a complete definition or by a partial
definition, i.e., a reductive statement in the narrower sense
of the word. There is no need here to enter into the details
connected with partial definitions; it suffices to say that such
a definition defines the meaning of a given term only in part,
i.e., it lays down the criteria for the applicability of a given
term only in reference to certain situations, leaving its applicability in other situations an open issue (possibly to be settled
by other partial definitions of that term).

Thus the reducibility, interpreted in this way, of terms of
a certain category (e.g., psychological ones) to those of another
category (e.g., physicalistic ones) does not ensure the full
translatability of statements which include terms of the former
category (e.g., psychological statements) into those which
include only terms of the latter category (e.g., physicalistic
statements). In this way the connection between the concept of
the unity of the language of science and that of the theoretical
unity of science was loosened; the former ceased to be a sufficient basis for the latter. In practice this meant the overthrow
of the concept of the theoretical unity of science. In a text
published three years after the paper of Hampel's under consideration Carnap wrote 'it is *obvious* [author's italics] that, at
the present time, laws of psychology and social science cannot
be derived from those of biology and physics.' What was left of
the original idea was the conviction that 'no scientific reason is
known for the assumption that such a deriviation should be in
principle and forever impossible', and the belief that 'the construction of one homogeneous system of laws for the whole
science is an aim for the future development of science'(23)
(though no explanation was given why that aim should be treated
as attractive).(24)

The fortunes of the concept of the unity of the language of
science took a different course. Being open to liberalization,
that conception kept its position in the doctrine of logical positivism - mainly in the circle of Carnap and his disciples - and
evolved together with the doctrine as a whole. That evolution,
however, resulted in a change in the nature of the concept,
which was transformed into an element of the concept of the
methodological unity of science.

In fact, the development of the concept of the unity of the
language of science had two stages. The first was marked by the
stand described above on the basis of Hempel's text, and,
originally formulated by Carnap in 'Die physicalische Sprache
als Universalsprache.'(25) The beginning of the second stage
was linked to the appearance of Carnap's 'Testability and Meaning',(26) which introduced the idea of reducibility. The ideas
formulated there led to Carnap's paper 'Logical Foundations of
the Unity of Science', already mentioned earlier, which was a
systematic exposition of the conception of the unity of the
language of science in its new, liberalized version. It was
expressed in the thesis that 'The class of observable thing-

predicates is a sufficient basis for the whole of the language
of science, including the cognitive part of everyday language',
which means that every extra-logical scientific term (and also
those terms which occur in the 'cognitive part' of everyday
language) is reducible to terms which denote observable properties of material objects.

The original thesis that all extra-logical scientific terms are
(fully) definable in physicalistic terms was thus modified in
two places: (i) 'definability' was replaced by 'reducibility',
(ii) 'the language of physics' was replaced by 'terms denoting
observable properties of material objects'. The first change
obviously means a liberalization of the original thesis. The
second one consists rather in giving more precision to the original thesis; in other words, it is a more adequate formulation
of the original intentions. Be that as it may, the new version
treated the issue of the unity of the language of science as a
problem of the property of the terminological base of science
and of its connections with the remaining scientific terms. This
offered an opportunity for abandoning the unfortunate idea of
reducing the language of one discipline to that of another, and
thereby for returning to the original problem: what conditions
the language of science should satisfy from the point of view of
the assumptions of empiricism. In a general way those conditions
among others had already been formulated in the 1920s, in what
was a kind of collective manifesto of the Vienna Circle:

> Since the meaning of every statement of science must be
> statable by reduction to a statement about the given, like-
> wise the meaning of any concept, whatever branch of
> science it may belong to, must be statable by step-wise
> reduction to other concepts, down to the concepts of the
> lowest level which refer directly to the given.(27)

Carnap's physicalism was a rather risky method of giving a
detailed interpretation of that thesis. In other words, the idea
of the unity of the language of science in general, and its
first, physicalistic version in particular, was a rather inci-
dental element, a kind of a by-product of the search for the
answer to the question: What makes the language of science
comply with the epistemological assumptions of empiricism? The
answer, which assumed the unity of the language of science, was
an unnecessary burden; it would have sufficed to specify the
appropriate conditions which scientific terms having empirical
sense must meet, without prejudice to the interrelations among
those terms, and especially to the issue of a common base of the
entire conceptual apparatus.

The discussions on the unity of the language of science evolved in just this direction: less and less was being said about
the unity of the language of science and about its common base,
and more and more about the language of science - in so far as
it is a language of science, and not of pseudo-science - having

definite methodological properties, in particular about its bearing a definite relation to immediate empirical data. It is another issue that those conceptions could usually be reformulated in accordance with the concept of the unity of the language of science. For instance, Carnap's 'The Methodological Character of Theoretical Concepts' does not raise the issue of the unity of the language of science explicitly, but the observation language Lo is so characterized that it may be treated as the common base of the language of science in accordance with the earlier explication of that concept.

Unlike the two preceding ones, the concept of the methodological unity of science is an indispensable element of the logical-positivist doctrine: not in the sense that logical positivism is unimaginable without that thesis, but in the sense that a system based on the epistemological assumptions of empiricism (and being in fact their expansion), with its practically boundless possibilities for ignoring what is really going on in science, must have implied that thesis (in the sense of de facto necessity): the tenet of the methodological differentiation of science would have to come from nowhere.

It is another matter that one does not have to resort to such an inner analysis of the doctrine to comprehend why the thesis of the methodological unity of science became an unshakeable dogma of logical positivism. First, it was a thesis adopted within the framework of positivist tradition.(28) Second, logical positivism was shaped as a trend intended to reform intellectual life by purifying it from all kinds of nonsense. The pride of place on the logical-positivist list of nonsense was held by metaphysics, but the philosophy of the social sciences, connected with the conception of Geisteswissenschaften, also ranked high. In practice, the concept of Geisteswissenschaften, and also other ideas which emphasized the methodologically distinctive nature of the social sciences were treated as special cases of metaphysical aberration. Defence of the idea of the methodological unity of science became one of the essential points in the program for combating metaphysics. 'Those who emphasize today the autonomy of the "humanistic social sciences" are metaphysically inclined and want to postulate a different sort of scientific activity for the social sciences,'(29) Schlick wrote, when expressing those associations which were to outlive the period of the antimetaphysical crusade and to retain the status of self-evident truth in the period of the 'positive analysis' of the methodological properties of science.

The thesis on the methodological unity of science assumed one of the two following forms, and that according to the context of analysis and the objectives of argumentation:

(i) description: there are no essential methodological differences among sciences;
(ii) recommendation: methodological differences among sciences must be removed.

The contradiction is merely apparent. The first thesis stated that all scientific disciplines, in so far as they are truly scientific, implement the logical-positivist model of science; the second stated the same in the language of a practical recommendation: if the existing proto-scientific and pseudo-scientific disciplines are to be raised to the level of true science, they must be cultivated in accordance with the logical-positivist model of science.

Neither of these theses was applicable to physics. As has been stated, the scientific nature of physics was not subject to evaluation, and any possible glaring discrepancy between the model and reality in this field had to be interpreted to the detriment of the model. In practice, the tenet of the methodological unity of science referred to the social sciences: it denied rational substantiation to the methodologically distinct nature of those sciences and caused all essential discrepancies between the binding methodological model and the state of things in those sciences to be interpreted as a manifestation of the retardation of the social sciences relative to physics. In this way the methodological model, which by assumption described the state of things in physics, could become the criterion of progress in the social sciences. As we shall see later, owing to this fact sociology found itself in a situation unknown to the natural sciences: it could easily settle the most basic controversies about its foundation.

The thesis on the methodological retardation of the social sciences was in harmony with current ideas about science, in particular with the belief that there is a disproportion in substantive advancement between natural science (primarily physics), on the one hand, and social science, on the other. That belief, which has never been subjected to the well-known test: 'What does it mean and how is it to be measured?', was a good starting point for additional argumentation in favour of cultivating the social sciences in accordance with the proposed methodological model. The arguments were as follows: the achievements of social science are very modest when compared to those of natural science, especially physics: this is due to the use of incorrectly chosen methods, hence the methods to be used should be chosen correctly, i.e., they should be those used in natural science, so that the research process will be in agreement with the methodological model proposed. This kind of pragmatic argumentation used to appeal more strongly to representatives of social science than did the doctrinal statements which boiled down to the claim that deviation from the methodological model meant deviation from science. In other words, representatives of the social sciences were more willing to accept the thesis on methodological retardation in the version which promised later successes than the version which undermined the scientific status of their respective disciplines.

The tenet of the methodological unity of science reduced publications on the methodological properties of the social sciences to a repetition of what had been earlier written on the methodological properties of the natural science, and to more or less ingenious interpretative operations intended to show that those cases which seemingly deviated from the proposed model in fact either agreed with that model or could be interpreted in terms of retardation. Let us see how that procedure was used in the now classical paper by Hempel on general laws and explanation in history.(30) This case deserves attention also because the conception of explanation (and the related ideas concerning prediction and theory construction) had an exceptionally strong impact upon sociology, being second in that respect only to the idea of making the conceptual apparatus more empirical.

Hempel's paper aimed at demonstrating that 'general laws have quite analogous functions in history and in the natural sciences',(31) but he was mainly concerned with explanation in history, and his thesis was that:

in history no less than in any other branch of empirical inquiry scientific explanation can be achieved *only* by means of suitable general hypotheses, or by theories, which are bodies of systematically related hypotheses.(32)

His definition of explanation is as follows:

The explanation of the occurrence of an event of some specific kind E at a certain place and time consists, as it is usually expressed, in indicating the causes or determining factors of E. Now the assertion that a set of events - say, of the kinds C_1, C_2, . . . C_n - have caused the event to be explained, amounts to the statement that, according to certain general laws, a set of events, of the kinds mentioned is regularly accompanied by an event of kind E.(33)

Explanation is thus defined in agreement with Humean tradition and corresponds to the now commonly known schema of deductive-nomological explanation:

$$L_1, L_2, \ldots L_r$$

Explanans

$$C_1, C_2, \ldots C_k$$

_____

E   Explanandum

where L_1, L_2, . . . L_r are general laws; C_1, C_2, . . . C_k form the description of the initial conditions, and the line indicates that the description of the fact to be explained (explanandum) follows logically from the set of general laws and the description of the initial condition (explanans).(34) Note also that the general laws are subject (by the definition of general law) to the condition of empirical verifiability, which, as we know from elsewhere, involves definite consequences when it comes to the extra-logical terms that occur in those laws.

It is further to be noted that the truth of Hempel's principal thesis is guaranteed by the definition of explanation (strictly speaking, by the definition of explanation and the concept of cause, which is the key concept in that definition): the problem 'can we in history (scientifically) explain facts otherwise than by reference to general laws?' here has the rank of the problem 'is a bachelor a married man?' The real question is: 'what is the relationship between explanation understood in this way and the actual research practice of historians?', and the substantive (as opposed to the definitional) part of Hempel's arguments pertains to that issue. Now Hempel admits that in fact historical publications do not usually conform to that model, but he thinks that the discrepancy is inessential in the sense of being eliminable:

> What the explanatory analyses of historical events offer is, then, in most cases not an explanation in one of the meanings developed above, but something that might be called an *explanation sketch*, Such a sketch consists of a more or less vague indication of the laws and initial conditions considered as relevant, and it needs 'filling out' in order to turn into a full-fledged explanation. This filling-out requires further empirical research, for which the sketch suggests the direction. ... The filling-out process required by an explanation sketch will, in general, assume the form of a gradually increasing precision of the formulation involved.(35)

Thus the discrepancy between the model of explanation and historians' research practice is due to the imperfections of the latter and should be gradually eliminated through improving historians' work in the direction indicated by the model. Thus we have here a diagnosis (retardation of historical research) and a research program (filling-out process) in relation to one of the basic functions of historical knowledge. Hempel eliminates the possibility of any other diagnosis by criticizing the conceptions whose names ('the method of empathetic understanding,' 'interpretation of historical phenomena,' 'the procedure of ascertaining the "meaning" of a given historical event,' etc.) suggest connections with the discussion of explanation by anti-positivists, but the said conceptions are arbitrary constructions of Hempel's. Hence Hempel's polemic can merely point to what the misunderstanding consists in. Here is an example:

> Closely related to explanation and understanding is the so-called *interpretation of historical phenomena* in terms of some particular approach or theory. The interpretations which are actually offered in history consist either in subsuming the phenomena in question under a scientific expla-

nation or explanation sketch; or in an attempt to subsume them under some general idea which is not amenable to any empirical test. In the former case, interpretation clearly is explanation by means of universal hypotheses; in the latter, it amounts to a pseudo-explanation which may have emotive appeal and evoke vivid pictorial associations, but which does not further our theoretical understanding of the phenomena under consideration.(36)

By ignoring the research practice of historians (his text includes one example drawn from a publication on economics(37) and one fictitious example) and the traditional philosophy of history, Hempel precluded the possibility of a matter-of-fact discussion of the problem of explanation in history. He was equally arbitrary in handling the problem of prediction in history. If his conceptions of explanation and predicting were consistently applied to social knowledge, and not to history alone, we would have to conclude that social knowledge does not offer any possibility of explanation and prediction. Hempel is obviously less rigorous on this point, which forces him to be inconsistent. Here is an example:

Analogous remarks [i.e., analogous to those made about the explanation sketch] apply to all historical explanation in terms of class struggle, economic or geographic conditions, vested interests of certain groups, the tendency to conspicuous consumption, etc.(38)

Now at least at the time when Hempel wrote his text, terms like, for instance, 'class struggle', did not have the kind of interpretation which would meet the logical-positivist conditions of being empirical, and hence 'an explanation in terms of class struggle' had to be unambiguously qualified - on the basis of the doctrine which Hempel propounded - as a pseudo-explanation or, to put it more precisely, as 'a pseudo-explanation sketch'.

Such examples abound in texts which analyze 'the logical structure of social science', and they well illustrate how one has to believe that the methodological model under consideration does not have merely a normative value in relation to that science.

The idea of the methodological unity of science is closely connected with the popular contrasting of 'naturalism versus anti-naturalism in the methodology of social sciences',(39) introduced by Popper in 'The Poverty of Historicism'. In Popper's opinion there are two schools in the methodology of the social sciences (his alternative term is 'the less successful sciences'):

According to their views on the applicability of the methods of physics, we may classify these schools as *pro-naturalistic*

or as anti-naturalistic; labelling them 'pro-naturalistic' or
'positive' if they favour the application of the methods of
physics to the social sciences, and 'anti-naturalistic' or
'negative' if they oppose the use of those methods.(40)

The popularity of this distinction has largely exceeded its
value. On closer examination those two opposing methodological
schools turn out to be two opposing schools of verbal behaviour.
Popper himself unintentionally draws our attention to this:

> Whether a student of methods upholds anti-naturalistic or
> pro-naturalistic doctrines, or whether he adopts a theory
> combining both kinds of doctrines, will largely depend on
> his views about the character of the science under consider-
> ation, and about the character of its subject-matter. *But the
> attitude he adopts will also depend on his view about the
> methods of physics. I believe this latter point to be the most
> important of all.*(41)

It is clear that, the naturalist is in favour of the application
of something which he calls 'the methods of physics' to the
social sciences. What methods then does he recommend using
in the social sciences? That we do not know; we only know that
he calls them 'the methods of physics', and people associate such
widely different things with that term that we can without much
exaggeration say that there are as many opinions as there
are naturalists, and that their opinions may or may not have
an essential common denominator. Exactly the same can, of
course, be said about the anti-naturalists. To make matters
worse, in some cases the views of a naturalist can be totally
convergent with the views of an anti-naturalist and at variance
with the views of another naturalist. This is why there is no
reason to treat 'naturalism' as a methodological category, unless
it occurs in the context of a conception of the methods of
physics.

Now as a rule, this is just what happens, which explains the
popularity of the distinction made by Popper. In Popper, the
opposition 'naturalism versus anti-naturalism' occurs in the
context of Popper's conception of the methods of physics. In
discussions among sociologists, that opposition usually occurs
in the context of some more or less vulgarized version of logical
positivism. This means that in practice a naturalist is an advocate of the use, in the social sciences, of methods which in the
doctrine of logical positivism are methods of physics, or, to put
it more generally, methods of natural science. In this way
naturalism acquires definite methodological features, but ceases
to be distinguishable from logical positivism in the methodology
of social science, and becomes a substitute name for that
methodology.

It is, however, a name with enormous persuasive power.
Sociologists have always been suspicious of philosophy and

reluctant to adopt ideas deriving from such a purely speculative discipline. Hence for a sociologist it is not a matter of indifference whether he declares himself in favour of the use of the methods of physics or for the acceptance of the recommendations of a certain philosophical doctrine, and hence it is not indifferent for him whether he is a naturalist or a logical positivist, even if the outcome is the same in practice. This is of course not to say that the distinction introduced by Popper explains the popularity of logical positivism in sociology. I merely claim that the distinction gave sociologists verbal assistance in assimilating logical positivism.

Among the leading representatives of logical positivism, only Otto Neurath can be considered familiar with social science; being an economist, and in a sense a sociologist, by training, he has a number of publications to his credit which may be classed as dealing with social science. There are, however, no grounds for assigning to him any significant status in the history of those disciplines. He was, as a scientist, an activist rather than a thinker, an organizer and popularizer rather than a researcher, and his texts often resembled political manifestos rather than scientific papers ('this view is rejected', 'liberation from metaphysics', 'he who accepts our line knows only statements about spatio-temporal objects: he is a physicalist').(42) The remaining leading representatives of logical positivism were either 'pure philosophers' or philosophers connected with such disciplines as logic, mathematics and physics. Generally speaking, the doctrine of logical positivism was a product of a milieu almost completely lacking in contacts with research practice in the social sciences. Note in this connection that the German anti-positivistic philosophy of the Geisteswissenschaften was a product of outstanding scholars in that field (Dilthey, Rickert, Windelband, Simmel and Max Weber) and was more an integral part of theoretical reflection than a metascientific doctrine. Windelband, when declaring 'the war against positivism' in his rector's speech of 1894, stood on his own ground, whereas the members of the Vienna Circle, when stating that 'the objects of history and economics are people, things and their arrangement', in their program of 1929, formulated an outsider's common sense opinion, an opinion which can hardly be denied, but which has no connection with the crux of the matter.

I do not think, however, that the logical-positivist system of views on the social sciences should be interpreted as a result of ignorance. It is rather a question of how the philosophy of science is pursued than one of actual knowledge. As has been repeatedly emphasized here, the logical-positivist conception of the philosophy of science offers opportunities for constructing methodological systems that are exceptionally resistant to confrontation with the facts. In practice, only a factor external to the system, namely the socially determined respect for

physics and physicists, causes the logical-positivist methodological doctrines to undergo changes, consisting, among other
things, in the elimination of what is glaringly at variance with
established facts. Such considerations of prestige do not work,
however, when it comes to the social sciences. Consequently
some version or other of 'the logic of the social sciences' is a
system totally resistant to criticism which changes only when
opinions of 'the logic of natural sciences' change. To mention
concrete cases, the essential point is, for instance, not the
fact that Hempel's text on explanation in history is based on a
superficial knowledge of the subject-matter, but the fact that
his way of pursuing the philosophy of science freed the author
from all need to carry out any research in that field. In the
same way Carnap, after every change of views on the language
of natural science, could extend his conclusions to the domain
of the social sciences without any need to verify whether this
was legitimate.

It is also worth emphasizing that in this respect the philosophy
of science has not essentially changed to this day. The struggle
against 'the metaphysical aberration' of the various conceptions
of the Geisteswissenschaften has ended in the social (though
not intellectual) defeat of those conceptions. Reinterpretation
of such systems like psychoanalysis and Marxism was abandoned
after a few endeavours.(43) Problems of the social sciences
were given a marginal place among the problems of the philosophy of science. But the basic system of views, shaped in the
1920s, 1930s and 1940s, has been preserved; it still functions
in the life of that discipline, or, rather, it reveals its existence
at those less and less frequent moments when there is a need
to adopt an attitude toward the social sciences.

Above all, the idea of the methodological unity of the language
of science and the companion idea of the retardation of the
social sciences still enjoy the status of self-evident truth. In
other words, statements on the social sciences are in fact reduced
to the claim that here everything is essentially the same as in
natural science, but on a lower level. It is also important that
even those trends which tend to make the philosophy of science
more empirical do not reveal any tendency to move beyond the
traditional system of stereotypes, for example, for Kuhn the
retardation of the social sciences is beyond dispute. Sociology
for him is a case of proto-science,(44) and it cannot be otherwise, because that is required by the uniform model of the evolution of science he has adopted, and by the resulting criteria
of the maturity of a scientific discipline.(45) The philosophy
of science was from the beginning constructed on a system of
beliefs that makes the very mention of a possible methodological
differentiation of science border on absurdity. As I tried to
show in chapter 2, Kuhn and the other representatives of the
historical orientation in the philosophy of science have made a
methodological, but not a theoretical, breakthrough in that
discipline. They are thus developing the philosophy of science

in the direction indicated by the traditional system of theoretical assumptions, which assumes the existence of a single
model of science and makes them seek it in physics. Even
though they advocate empirical studies of the methodological
foundations of natural science, their road to analogous studies
of the social sciences is blocked.

5 From the methodological doctrine to research practice

'The methodological character of theoretical concepts' by Carnap, which introduced another essential liberalization of logical-positivist requirements concerning the conceptual apparatus of science, includes a passage which is an excellent illustration of the nature of the connections between philosophy of science and the social sciences:

> In psychology still more than in physics, the warnings by empiricists and operationists against certain concepts, for which no sufficiently clear rules of use were given, were necessary and useful. On the other hand, perhaps due to the too narrow limitations of the earlier principles of empiricism and operationism, some psychologists became over-cautious in the formation of new concepts. Others, whose methodological superego was fortunately not strong enough to restrain them, *dared to transgress the accepted limits*, but felt uneasy about it. Some of my psychologist friends think that we empiricists are responsible for too narrow restrictions applied by psychologists. Perhaps they over-estimate the influence that philosophers have on scientists in general; but maybe we should plead guilty to some extent. All the more should we now emphasize the changed conception which *gives much more freedom to the working scientist in the choice of his conceptual tools*.(1)

The philosophy of science is in this context a normative system that implies the rules of research procedure, and the philosopher of science, qua an expert in that field, is the scientist par excellence (a scientist of a higher order?), the methodological guru of the 'working scientists'.

We see here a divergence between two conceptions of the philosophy of science: one laid down in programmatic texts, and another actually practised. The former excludes interference with the research practice of the 'working scientists' by treating the philosophy of science either as a descriptive discipline or as a normative system based on a certain convention and hence binding only within that convention. Carnap in principle engaged in descriptive methodology, while Popper presented his system as one of rules which are subordinated to a certain conventionally adopted primary rule (by means of which he defines science). Methodology treated in this way may be useful in comprehending science (Popper: 'My only reason for proposing my criterion of demarcation is that it is fruitful: that a great many points can be clarified and explained with its help'), but is not a system that would bind the researcher, which is stated explicitly ('there are no recipes for the pursuit of science', 'the philosopher is not competent to teach a resear-cher who is an expert in his field', etc.). In fact, however, philosophers of science grant their systems a two-fold status, both as a pertinent description of science and the source of universally binding norms of what is scientific.

This is quite comprehensible: every conception of science, rationality, beauty, justice and other similar goods of a higher order has the essential properties of a revealed truth. As such it both informs and instructs, and the author of such a con-ception, regardless of what he may declare, has a proselyte's approach to what in his interpretation is scientific, rational, etc. A normative system need not be (and usually is not) formulated in a language which in philosophical interpretation is considered a language of norms. Thus in practice the philosopher of science, in agreement with his declaration, does not instruct the researcher on how the latter should proceed, but only says how one proceeds in science; he does not require that the researcher adjust the results of his activity to definite standards, but merely informs what standards are obligatory in science; he does not interfere with research techniques, but often has to state that they are pseudo-scientific.(2)

The impact of the philosophy of science upon other disciplines depends, however, not so much upon the intentions and aspira-tions of the philosophers as upon the attitude of the recipients; not so much upon the role which the philosophers assign to their doctrines as upon the attitude toward those doctrines on the part of representatives of the various academic disciplines. In that respect the philosophy of science found itself in a seemingly paradoxical situation. Its interests and aspirations were (and are) connected with physics and related disciplines; its successes, with sociology and related disciplines. The logical analysis of physics was, without exaggeration, a scienti-fic revelation among sociologists without greatly impressing physicists. The philosophical model of science, in its distorted and vulgarized form, became one of the fundamental elements of the intellectual equipment of sociologists, and functions, in the main trend of sociology, as its system of basic theoretical and methodological assumptions. In the present chapter we shall be concerned with a general description of this form of philosophy of science, usually not known to philosophers, and its role in the research work of the sociologists.

'The methods of science have been enormously successful wher-ever they have been tried. Let us then apply them to human affairs.'(3) This statement comes from B.F. Skinner, one of the principal figures in behaviouristic psychology and sociology, but it could come from at least every second contemporary

sociologist, because that statement expresses the belief which
is stereotypically repeated in texts which advocate the use of
'the methods of natural science' in sociology. Obviously Skinner
assumes (i) the existence of something like (specific) methods
of science or, as others say, scientific methods (or even just
'the scientific method'); (ii) that these methods constitute a kind
of instrument that can be used in various situations and in the
study of various objects; (iii) that the successes of natural
science are connected with the use of that instrument. Under
such assumptions the requirement of introducing 'the methods
of science', i.e., in fact, 'the methods of natural science', at
least sounds reasonable and one may suspect in advance that
the opposite standpoint is based on irrational assumptions
(Lundberg thought it was a matter of obscurantism and prejudice).(4)

The idea of healing the human sciences by the 'methods of
science' is, of course, not new. For over a hundred years (so
as not to go still further) it has been the point of departure
for more or less serious methodological programs, not necessarily connected with positivism. Skinner's requirement follows
from the same assumptions and voices the same hopes as Comte's
programme of positive sociology, Durkheim's demand that social
phenomena be treated like objects, Radcliffe-Brown's proposal
to found 'the natural science of society',(5) the programme of
'scientific sociology' from the 1940s, and such undertakings
as 'the barometer of international security'.(6) The belief that
the human sciences must be 'made scientific' by means of the
'methods of science' brought together, as the above examples
show, people so different in their intellectual profiles, that the
real importance of that common tie must appear suspect. In
fact, those 'methods of science' turn out to be a variable for
which the most diverse, sometimes extremely naive, things are
substituted. It can easily be guessed that this is due not to an
inclination to adopt different interpretations, but to superficial
knowledge of natural science.

The attitude of social scientists to natural science has always
resembled love of a beautiful stranger, a love which, as usual
in such cases, abounds in comical elements. The year 1942 saw
the appearance of a bulky book (of nearly a thousand pages)
by Stuart C. Dodd, entitled 'Dimensions of Society', whose
intention was to make sociology resemble natural science, and
which latter its author unambiguously associated with 'mathe-
matization', i.e., with the use of symbolism. The mathematician
who reviewed the book states that there is no mathematics in
it, but a 'reckless abuse of the mathematical vocabulary'; that
problems such as 'Can dimensional analysis of societal situations
be used as dimensional analysis is used in physics...?' have as
much sense as the problems like 'How many yards of buttermilk
does it take to make a pair of breeches for a bull?', and that
it may impress the layman, but the mathematician sees in it
merely a large dose of pretentiousness.(7) The review ends

with a statement which, I believe applies to the whole 'naturalist' undertaking in sociology:

> Only sociologists can say whether this is what they want.
> If it is, the book offers no hint of how to produce theory,
> desirable or otherwise as it might be, beyond repeated
> exhortations similar to those which urge the faithful to get
> to Heaven if they can.(8)

For the sake of completeness, note that in 'Dimensions of
Society' there is no sociology either.(9) More precisely, the
book includes some extremely primitive social knowledge, based
on current stereotypes and prepared so as to suit that language
of Dodd's 'theory' (which he gave the 'scientific' name of
'S-Theory').

If the development of knowledge followed any of the known
philosophical conceptions of rational cognition, works like
'Dimensions of Society' would immediately be eliminated from
scientific life as evident failures. Yet the 1940s and 1950s
brought a marked increase in such sociological production,
accompanied by gradually dwindling criticism. We can mention
here works referring substantively to physics, like Kurt
Lewin's notorious 'field theory', and also those referring
methodologically to physics like 'measurements of attitudes',
based on the 'scales' of Bogardus, Likert, Guttman, and
Thurston; they evidently failed to comply with formal requirements and were founded on naive substantive assumptions.
This whole development cannot be explained without reference
to the social mechanisms of selections in science.(10)

In light of the above, it is easy to understand the role
methodology has played in sociology during the last fifty years.
It became the link between true science and sociology, an
interpretation of the scientific method that was both professional and accessible to non-professionals. Methodology thereby
not only provided an additional and deeper justification of the
pro-naturalist tendencies, but also practical possibilities for
putting them into effect. Comte, when advocating his programme
for the positive science of society, based it on his own (ama-
teurish) inquiries into the nature of the positive sciences; the
contemporary sociologist who advocates pursuing sociology
according to the pattern of the natural science has that pattern
at hand, owing to handbooks of methodology.

In order to avoid misunderstandings it is worth while introducing a distinction between internal and external methodology
in sociology. The terms may sound somewhat artificial, but
they are used here to emphasize the most essential features of
these two kinds of activity.

Internal methodology has its source in the research practice
of sociologists and articulates that practice; external methodology comes from systems connected with other disciplines
(philosophy, logic, mathematics). Thus internal methodology

forms an integral part of sociology and is pursued as a self-
reflection on the part of practising sociologists: methodological
considerations and discussions are forced upon them by substantive problems and are closely linked to those problems,
which accounts for the fact that dividing the sociologists into
researchers and methodologists (and even singling out the
category of methodologists) is as absurd as a division into
researchers and theorists. The external methodology has its
roots in the problems of its original discipline, and its pursuit
requires expert knowledge which the sociologist usually lacks
and which he can acquire only in a very limited sphere. If the
methodology is taken to mean external methodology, the division
into methodologists and researchers is a necessity.(11)

The pivotal place in internal methodology is held, of course,
by technical issues,(12) but these cannot in any way be
identified with the methods of field studies, since discussions
on technical problems are usually accompanied by reflections
on the theoretical and epistemological foundations of research
techniques. It suffices to recall here the discussion of the
value of personal documents in social research and publications
by such authors as Charles Cooley, Robert MacIver, Florian
Znaniecki, Charles Ellwood, and - among the contempories -
Howard Becker and Aaron Cicourel.(13)

Internal methodology is a natural element of every academic
discipline, even though it is sometimes difficult to single out
the methodological contributions of the various sciences.
External methodology singles out sociology and related disciplines (especially psychology and political science) among other
academic disciplines and is a result of the search for 'the
scientific method' outside one's own field.

What has ultimately taken shape as methodological specialization within sociology is a system which is completely dominated
by external methodology and in this sense reflects the philosophy of science and some branches of applied mathematics.
The links between that system and sociological research techniques are fairly strong; comparison of methodological works
by such authors as Paul Lazarsfeld, Hans Zetterberg, Stefan
Nowak, Raymond Boudon and Hubert Blalock with what Hempel
and Nagel have written on the social sciences enables us to
understand the difference between real methodology, pursued
by people who know their research techniques, and armchair
methodology. This does not, however, change the fact that the
foundations of the 'methodology of social research'(14) remain
outside sociology and are a product of thinking which in fact
had no connections with sociological thinking. The issue will
be discussed later in greater detail; for the time being let us
note that what is meant here is not more or less essential borrowing, but the nature of the entire undertaking: the 'methodology of social research' has become a conveyor belt which trans-
mits the knowledge of 'the scientific method', i.e., in fact, the
methodological ideas of those authors who pass for experts in

'the scientific method' to sociologists (Carnap's 'working scientists').

In that role the 'methodology of social research' soon became the strongest, the most prestigious and the most influential speciality, with its own organizational forms (sections of sociological societies), its own periodicals (e.g., 'Sociological Methods and Research'), its own serial publications (e.g., 'Sociological Methodology') and, especially in the United States, a greatly expanded curriculum.(15) Its impact on sociology proper is based, among other things, on the separation of research techniques from the process of research. This is very common among sociologists: the process of research is conceived as an operation consisting in the application of research techniques in a definite research situation (solution of a definite research problem). 'The purpose of research is to discover answers to questions through the application of scientific procedure', say the authors of the handbook on which several generations of sociologists have been educated.(16) Such assumptions easily result in treating sociology as a kind of applied methodology of social research. David Willer (who by the way sharply opposes what he calls positivism and imitation of natural science in sociology) thinks that sociology is not a science (although it should be) and that the reason for that is the lack of an adequate methodology of sociology, and in particular a methodology of constructing sociological theories:

> How is it possible for theory to be constructed in sociology *before* a methodology for constructing theory exists? Even if sociologists wished to construct testable theory today, there are no established *guidelines* for its construction. ... It seems clear that tested theory will not arise spontaneously without the *prior* existence of a methodology for constructing that theory.(17)

Willer went to the extreme, but the belief that the pursuit of sociology consists in applying methodology in practice is common. The average sociologist sees the relation between sociology and methodology as Lakatos saw the relation between history and the philosophy of science: sociology without methodology is blind. This obviously gives the 'methodology of social research' the rank of a system theoretically superior to sociology proper, and the methodologists, i.e., representatives of that subdiscipline, the status of sociologists par excellence. An indication of that status is the fact that, as Edgar Borgatta noted, sociology departments have 'a propensity to grant PhD's to students who had achieved the competence of an undergraduate major in mathematics, independently of sociological knowledge.'(18) Borgatta has American Universities in mind, but the phenomenon is not confined to the United States. Let us add that a PhD in sociology can equally well be granted for a thesis which expounds the standard logic of

science in an appropriate language version ('the structure of
sociological theory' instead of 'the structure of theory',
'nomological explanation in sociology' instead of 'nomological
explanation', 'levels of measurement in sociology' instead of
'levels of measurement', etc.), with the reservation that
such an exposition rarely comes up to the standard of an
advanced handbook on the philosophy of science. All this
obviously applies not to doctoral dissertations only: as has
been mentioned, the 'methodology of social research' is a derivative undertaking, a kind of philosophy of science plus
mathematics for sociologists, and such undertakings usually
do not equal their models.

The thesis on the essential dependence of the 'methodology
of social research', and hence also the main current of contemporary sociology, upon the philosophy of science is not
undermined by the lack of significant working contacts, not to
say co-operation, between the representatives of these two
disciplines, nor by striking differences in research techniques,
nor by substantive divergences. George A. Lundberg did not
know the publications of the Vienna Circle,(19) Paul Lazarsfeld
blamed the philosophers of science for ignoring the real
methodological problems of the social sciences,(20) and Hans
Zetterberg, when writing about the axiomatization and the
empirical verification of sociological theories, almost exclusively
quoted sociologists and psychologists, and his knowledge of
the texts on that subject by philosophers of science was
extremely modest.(21) Yet each of these authors is commonly
included in the group of leading representatives of positivist
sociology, which is fully justified by, among other things, the
close connection between their respective methodological opinions, on the one hand, and logical-positivist methodology, on
the other hand (although those connections are of a different
nature in each of these three cases). The movement of ideas
does not necessarily follow the rules of standard content analysis. In sociology logical positivism played the role of a Weltan-
schauung. Its first followers were confessors of, rather than
experts in, 'scientific philosophy'; they easily assimilated its
general methodological slogans, but later followed their own
course, being guided by their common sense rather than by
expert philosophical knowledge.

This gave rise to a specific sociological variety of the logical-
positivist methodological doctrine, extremely orthodox although
at first glance giving the impression of a relatively independent
creation. This impression was strengthened by the coinciding
of that doctrine with current epistemological conceptions and
the current knowledge of natural science.

As time passed, the principal methodological assumptions of
logical positivism were more and more treated as an element of
the anonymous 'achievement of science'. It is this fact, and
not any uncontrolled tendency to plagiarize, which explains why
the protagonists of the new trend, not to speak of lesser

figures, did not mention the authors of the views they pro-
pagated as frequently as is customary. When Skinner called for
the application of 'methods of science' in social research, he
did not have to explain what he meant, nor did he risk his
slogan being interpreted as an invitation to investigate
research techniques in physics. The methods of science were
given. There was also a vast body of knowledge on the subject
and a group of recognized experts in the field which, of
course, included philosophers of science, and not, for instance,
physicists. The latter, like sociologists, were ranked among
working scientists, that is those who apply the methods of
science, but do not necessarily realize what they are doing,
even if they do it well. The acceptance of a pro-naturalist slogan,
the acceptance of the requirement that sociology should be made
scientific, thus implied the necessity of sharing in that knowledge; this was synonymous with mastering the foundations of
research methods, and thus with elementary professional training. Hence the role of methodological initiation as the indicator
of professional perfection, hence the expanded and largely
institutionalized system of introducing methodological knowledge
in the process of research (ordinary university courses, courses
in special methodological training, systems of methodological
consultation, the methodological 'quality testing' of research),
hence also the above-mentioned status of the methodological
speciality within sociology.

I keep returning to that issue and emphasize it in various
contexts because it seems that it is there that we must look for
the key to understanding the fact that sociology, a discipline
with its own traditions, its own intellectual profile and considerable achievements, has become dominated and basically
transformed by ephemeral conceptions from the outside. This
fact, unprecedented in the history of science, can of course,
be interpreted in different ways, which is an undertaking of
extreme interest for the sociology of knowledge. But here we
are interested in the explanation in terms of the internal
history of academic thought. From that point of view, the
decisive role was played by sociologists' pro-naturalist sentiments and their metaphysical approach to 'the scientific
method'. Without belief in the existence of the scientific method
as such, the requirement of making sociology scientific by the
application of that method would not make sense (as there
would be no sense in the concept of research as 'the search
for answers to questions by the application of the scientific
method'). Their insistence that there was such a thing as
'the scientific method' did not settle the issue of where that
method was to be sought nor who was to be treated as its
spokesman, and hence did not yet lead directly to an uncritical
identification of 'the scientific method' with what was presented
as the scientific method by the philosophy of science, but
it was conducive to a separation of the knowledge of methods
from the knowledge of science, and to the acceptance of

methodological reflection pursued separately from scientific reflection as the source of knowledge of the methods of science. Owing to that, Carnap, Hempel and Nagel appeared as professionals, and not as philosophers, and their texts were treated as sociologists would treat a competent text on microbiology or mechanics, that is, as the authoritative source of knowledge in a definite field.

This is worth emphasizing: in the opinion of sociologists the conceptions of the philosophy of science were not associated with philosophy, logical positivism, empiricism, or anything that would suggest that those conceptions were represented merely a certain opinion, a certain view point, etc. They were treated as the source of 'positive knowledge'. They were and still are treated as such. It is true that now people notice the 'relativity' of the various methodological conceptions of the philosophy of science, but only in the sense that this discipline is no longer treated as a monolith shaped in the process of simple cumulation of knowledge; it is clear now that it is marked by inner differentiation, rival systems of opinions, the collapse of some conceptions, etc. Thus, for instance, the faith in Carnap-like inductionism did collapse, but the faith in the philosophy of science as the authoritative source of knowledge of scientific methods did not. There is a kind of modernization of imports: Carnap's conceptions have been replaced by those of Popper, and these in turn by those of Kuhn. If a glaring example is needed to show that the methodological thinking of sociologists is that of epigones, it can be found in the career of Kuhn's ideas, and in particular of the concept of 'paradigm' in the latest history of sociology. This is what Feyerabend notes on this subject:

> More than one social scientist has pointed out to me that now at least he had learned how to turn his field into a 'science' - by which of course he meant that he had learned how to *improve* it. The recipe, according to these people, is to restrict criticism, to reduce the number of comprehensive theories to one, ad to create a normal science that has this one theory as its paradigm. Students must be prevented from speculating along different lines and the more restless colleagues must be made to conform and 'do serious work'.(22)

Feyerabend colors, but does not exaggerate. For many present-day representatives of the social sciences, especially for sociologists, the 'paradigm' is what the 'operational definition' was for their predecessors, namely an intellectual revelation. It is a fact that that concept happens to be used in a much better way than would follow from the above quotation. As an example one can mention 'A Sociology of Sociology', by Robert W. Friedrichs.(23) But did sociology, which has the sociology of knowledge to its credit, have to wait for a signal from the philosophy of science to notice the social nature of the mechanisms that explain the evolution of scientific thought, the more so as the point was to explain the evolution of social scientific thought?(24)

To sum up, we have to say that the logical positivist philosophy of science in sociology found an utterly uncritical partner, a partner who expected methodological instructions and accepted them as the binding norm, regardless of the intentions of the instructor. The crisis in the philosophy of science has brought no change in that respect. A contemporary sociologist who specializes in methodology, when addressing a group of philosophers of science in 1965, said in conclusion:

> I think it was Dr. Lakatos whom I heard characterize the work of philosophers of science as telling scientists what they did and why they did it after they had done it. Perhaps *some of you should perform these same functions for sociological theorists before they do the kind of theorizing that they are at present doing.* In suggesting this, however, I am not unaware of the fact ... that philosophers of science themselves appear to be divided on the kind of *advice and counsel they would give us.*(25)

George C. Homans, when recalling his evolution as a social scientist wrote: 'Following up my originally rather inchoate dissatisfaction with grand theory, I took the trouble to read what the philosophers had been recently writing about the nature of scientific theory.' He means such philosophers as Braithwaite, Nagel and Hempel. The reading of their works made him develop the following theory of theory:

> A theory of a phenomenon is an explanation of it. An explanation consists of at least three propositions, each, including the proposition to be explained (*explicandum*), stating a relationship between properties of nature. The propositions form a deductive system, such that the *explicandum* follows as a conclusion in logic from the others in the set: this is where logic comes in. At least one of the propositions has to be more general than the others: hence this is known as the 'covering law' view of explanation.(26)

To put it briefly, a theory consists of at least three propositions which state relationship between properties of nature and together form a 'deductive system' (it seems that for Homans every set of propositions connected by the relation of logical inference is a 'deductive system'). Sociological theories (i.e., that which traditionally passes for sociological theories) usually consist, as we know, of a great many statements each, and never (Homans would say: almost never) have the properties of a deductive system; they also do not meet the conditions imposed by Hempel upon a system of explanatory statements, nor are they a 'deductive system' in Homans's sense, whatever

that should mean. This makes Homans formulate a very unfavorable diagnosis for the social sciences in general, and sociology
in particular: they are disciplines without theoretical achievements, or rather in a pre-theoretical stage (in view of the fact
that advances in behavioural psychology bode well for the
future).

Homans is a well-known reductionist and behaviourist, but
he does not appear here in either of these roles. Behaviourism
and reductionism occupy little place in the present book because
they are relatively independent of the logical-positivist methodological system despite the fact - but that is another issue -
that they are largely convergent with that system.(27) Homans
could expound his reductionist and behaviourist views without
seeking support for them in the conceptions of the philosophy
of science, but what he says on the nature and structure of
sociological theories and the evaluations which, as a result,
he formulates concerning the theoretical achievements of the
social sciences as a whole, come from Hempel, Braithwaite and
perhaps a few others philosophers of science.

Of course, it is not Hempel who claims that 'the explanation
of a phenomenon is the theory of the phenomenon: no other
meaning can be given to the word "theory".'(28) Homans
distorts the logical-positivist conception of theory and completely trivializes Hempel's schema of explanation by ignoring
everything which complicates the original version of that
schema, formulated in the 1940s. Hempel as seen in Homans's
texts is a caricature of Hempel as seen in Hempel's texts,
especially the later ones, such as 'The Theoretician's Dilemma'.
Also significant is the fact that Homans does not understand
many key concepts in the philosophy of science (such as those
of a deductive system and of a general law)(29) and, by using
his linguistic intuition, he constructs his own coherent and
closed system of methodological concepts. All this speaks in
favour of exonerating Hempel, or anyone else, from responsibility for the views expounded by Homans, but this does not
alter the self-evident fact that those views are copies of
corresponding parts of the philosophy of science. True, these
copies are incomparably worse than the original, but this is
just one of the principal problems with which we are concerned
here. The point is not to establish who is responsible for the
methodological profile of the main trend in contemporary sociology, but to trace the fortunes, or misfortunes, of certain
methodological ideas, and to demonstrate how those ideas came
to affect research practice in sociology. The distortions and
trivialization of those ideas in the process of their adaptation
are an integral part of that process.

Homans uses his 'scientific' theory of theories in a manner
which is typical of the trend here under consideration: he
strives to demonstrate that every theoretical orientation which
competes with his own is of necessity pseudo-scientific or
pre-scientific. 'The weakness, here and now, of those opposing

the position taken here is that, while arguing in general terms that psychology will never explain social phenomena [who claimed that?], they do not produce alternative kinds of explanation.' Is this really so? Of course they do not produce them - says Homans - because explanation consists in constructing an appropriate deductive system. Therefore,

> let them take the problem of explanation seriously. Specifically, let them stop citing in their favor the facts of social emergence, social constraint, even of a social whole's being greater than the sum of its parts ... . Let them rather take specific phenomena that, in their view, exemplify emergence, constraint, wholeness, or indeed anything else, and show how they would explain them *sketching out their deductive system in some detail.*(30)

This amounts to saying: Do as I do; that will best show that you do not do it differently. It is obvious that the protagonists of methodological holism (Homans uses the term 'methodological socialism'), functionalism, historicism, etc., are unable to construct any 'deductive system', and hence are not in a position to 'explain' anything; it follows that there is no 'theory' in their works. But what is there? Now, according to Homans, the output is mainly of two kinds: 'nonoperating definitions' and 'orienting statements'. The description of the concept of 'non-operating definition' makes us conclude that he means something like 'theoretical definitions' (as that term is used by Zetterberg, for instance), that is, explication of the meaning of a given term by means of theoretical terms understood as (extra-logical) non-observational terms. In opposition to 'non-operating definitions' we have 'operating definitions', exemplified by the definition of pressure which accompanies (as Homans puts it) Boyle's law. As examples of 'non-operating definitions' Homans mentions definitions of such traditional sociological concepts as role and culture. 'Orienting statements' are those which look like 'real propositions' but really are not. That category includes some of the most renowned statements in the social sciences. Homans' pet example is 'Marx's statement that the organization of the means of production determines the other features of a society.' Homans comments on it as follows.

> This is more than a definition and resembles a proposition in that it relates two phenomena to one another. But these phenomena - the means of production and the other features of a society - are not single variables. At best they are whole clusters of undefined variables. And the relationship between the phenomena is unspecified, except that the main direction of causation - determination - is from the former to the latter. Whereas Boyle's Law says that, if pressure goes up, volume will assuredly go down, what Marx's Law says is that, if there is some, any, change in

the means of production, there will be some unspecified change or changes in the other feature of society. Put the matter another way: Boyle will allow one to predict *what* will happen; Marx will only allow one to predict that *something* will happen. Accordingly I cannot grant his law the status of real proposition.(31)

It must be emphasized that this evaluation is based on the merits of the theory he analyzes, and is not, for instance, a rationalization of an ideological evaluation. Whatever Homans would think of Marx in general(32) his appraisal of the above thesis would be the same. He assigns, e.g., 'Toward a General Theory of Action' by Parsons and Shils, to the same group. In Homans's opinion, the world of intellectual products, at least those which result from academic activity, falls into two parts: one consists of non-operating, i.e., in fact apparent, definitions and orienting statements (it seems, although this is not quite clear, that orienting statements are marked by the occurrence in them of terms introduced by means of non-operating definitions); the other consists of operating definitions and real propositions (by analogy, real propositions seem to be marked by the fact that the extra-logical terms which occur in them are introduced by means of operating definitions). In Homans's opinion, the former part includes an overwhelming majority of the achievements of the social sciences.(33)
Homans by no means thinks that he proves his claims by resorting to appropriate definitional operations:

> Naturally any scholar is free to use the word 'theory' in any way he likes, even for something different from what I call theory, provided he makes clear just how he is using it and does not, by slurring over the issue, claim for his kind of theory, by implication, virtues that belong to a different kind. All I submit here is that, *normally in science*, 'theory' refers to the sort of thing I have described.(34)

But on another occasion he seems to take a different standpoint:

> The position I have just taken as to the nature of explanation and of theory is now fairly well accepted by the philosophers of science. It ought to be familiar to social scientists, and indeed many of them would accept it if it were put to them in the abstract. *But in what they actually say, and in the sort of thing they in fact call theory, they often appear to hold views about the nature of theory that are far different from one stated above.* This is particularly true of sociology.(35)

How are these two statements to be reconciled? On the basis of the same logic which I tried to reconstruct in chapter 4 when analyzing the attitude of logical-positivism towards the social sciences:(36) it is said that the suggested conception of theory is a model of what is accepted as theory in science, then does the conflict between that model and that which is accepted as theory in the social sciences undermine the model? No, because what is accepted as theory in the social sciences does not meet the (model) requirements of a theory. Is that then a normative model? No, because it merely reconstructs that which really functions as theories in science. Does it then apply to only some disciplines? No, because it is a model of scientific theory in general.

Such reasoning is to be found in the otherwise best studies of the subject. Of course, in practice, the ideas follow a much wider course, but the result is the same. We are dealing here less with a specific style of thinking than with thinking in a specific theoretical context.

A characteristic feature of positivism in general, and logical positivism in particular, is the regulation of thinking by the imposition of norms and patterns that control intellectual output - mainly, though not exclusively, in those fields tradition-ally connected with academic institutions. The doctrine of logical positivism is, in fact, a complex and quite coherent normative system that implies specified answers to a large number of questions, from epistemological to purely technical ones. That system, like other living normative systems, is given in a descriptive account, but, also as in the case of other systems of that kind, the requirements of description are subordinated to those of persuasion. The doctrine of logical positivism is by assumption the explication of the methodological nature of science. Hence interest is focused not so much on actual scientific activity as on the ideal entity which manifests itself in actual scientific activity. Manifestations of ideal entities are, by their very nature, more or less imperfect, the degree of their perfection being measured by their agreement with the ideal. This may mean that manifestations 'are allowed' to deviate from the ideal, but it may equally mean that the ideal 'is allowed' to deviate from manifestations: everything depends on the direction in which the measure is applied. In any case, this offers opportunities for a fairly free manipulation of the empirical data, which can (though certainly need not - this is a matter of the researcher's attitude) result in making a con-struction that is resistant to empirical criticism. The logical-positivist model of science is just such a construction. The possibility of interpreting what happens in science here and there as more or less imperfect manifestations of the nature of science is used in their model to endow that ideal entity with all those attributes which science should have on the basis of the adopted system of norms. In this way the description has been transformed into an explication of the norm: instead of writing about what methodological properties science should have people write about what properties science (*as such*) does

have. Thus, Homans's statements above quoted include no contradictions. Some of them speak about the spirit of science,
i.e., about how things should be in science, others speak about
some deplorable manifestations of that spirit, i.e., about what
takes place in the social sciences. This confusion of entities
is still clearly manifested in many meta-sociological publications.

Homans is a comparatively modest and little known (in that
role) representative of the tradition initiated by the renowned
and very influential book by Hans Zetterberg, 'On Theory and
Verification in Sociology'.(37) Zetterberg based himself on
C. West Churchman's 'Elements of Logic and Formal Science'
to present his 'axiomatic theory', which he also termed a
'deductive-type theory'; he postulated construction of such
theories in sociology because they form 'the most efficient form
of formulation *known to the logician*.'(38) Zetterberg characterized the 'axiomatic theory' by describing the procedure that
leads to its formulation. Now

> if we were to formulate an axiomatic theory – or deductive-
> type theory, as it is also called – we would proceed in
> roughly the following way.
>   1 We will list a series of primitive terms or *basic concepts*.
> These are definitions which we, strictly speaking, will intro-
> duce as undefined. ...
>   2 We will define *derived concepts* of our theory by means
> of these concepts. The derived concepts are obtained through
> combination of the basic concepts. ... The basic concepts
> and the derived concepts form the 'nominal definitions' of
> the theory.
>   3 We will formulate the hypotheses of the theory. These
> hypotheses must not contain any other concepts than the
> nominal definitions.
>   4 We will select from among the hypotheses formulated a
> certain number to be the *postulates* of our theory. The
> postulates should be chosen so that all other hypotheses,
> the *theorems*, should be capable of derivation from these
> postulates.(39)

If we disregard small inexactitudes (e.g., confusion of
'concepts' with 'terms' and 'terms' with 'definitions') we can
easily note that 'an axiomatic theory' in Zetterberg's sense is
obtained from a combination of a certain number of simple
(basic) elements, which can be manipulated in various manners
so that they yield different wholes. Zetterberg, in fact, was
fascinated by the possibility of logical manipulation of propositions, and in particular with combining them into coherent
wholes, reducing a set of propositions to one of its subsets,
etc. This was because he believed that owing to a large number
of empirical studies in sociology, there were many well substantiated but not interconnected propositions, and that the

task of a theoretically-oriented sociologist was to combine those propositions into more comprehensive wholes. Thus the formation of 'axiomatic theories' for Zetterberg was primarily the process of ordering and systematization of verified sociological propositions. On the other hand, he pointed to the advantages of such systematization for the further process of verification of sociological propositions.

This was thus the outline of a certain research strategy. Zetterberg stated plainly that sociology had not yet attained the level of a discipline that has its own axiomatic systems, although it was already in a position to have them. To illustrate the second part of his claim he used the example constructed by himself, which we must unfortunately quote in its entirety because only that example makes it possible to fully comprehend what an 'axiomatic theory' in sociology is and how it is constructed.

Let us suppose that we have studied a number of groups with respect to: (a) the number of members in the group; (b) the solidarity of the group; (c) the degree of uniformity of behavior around the group norms; (d) the degree of deviation from group norms; (e) the division of labor in the groups; and (f) the extent to which persons are rejected or pushed out of the group when they violate group norms. These six variables, we assume, are found to be interrelated in the following way:

1. The greater the division of labor, the more the uniformity.
2. The greater the solidarity, the greater the number of members.
3. The greater the number of members, the less the deviation.
4. The more the uniformity, the less the rejection of deviates.
5. The greater the division of labor, the greater the solidarity.
6. The greater the number of members, the less the rejection of deviates.
7. The greater the solidarity, the more the uniformity.
8. The greater the number of members, the greater the division of labor.
9. The greater the division of labor, the less the deviation.
10. The less the deviation, the less the rejection of deviates.
11. The greater the solidarity, the less the rejection of deviates.
12. The greater the number of members, the more the uniformity.
13. The greater the division of labor, the less the rejection of deviates.

14   The greater the solidarity, the less the rejection of
deviates.
Summarizing those findings according to the model of the
axiomatic theory we can proceed like this:
Introduce as basic concepts:

behavior	solidarity
member	division of labor
group	rejection
norm	

Formulate as derived concepts:

uniformity:	the proportion of members whose behavior is the norm of the group;
deviation:	the proportion of members whose behavior is not the norm of the group;
deviate:	member whose behavior is not the norm of the group.

Select as postulates, for example, findings (5), (7), (8),
and (14):
I. The greater the division of labor, the greater the
solidarity.
II. The greater the solidarity, the more the uniformity.
III. The greater the number of members, the greater the
division of labor.
IV. The greater the solidarity, the less the rejection of
deviates.
It is easily realized that from these four postulates all
fourteen findings can be derived as they are combined with
each other and with the nominal definitions.(40)

Zetterberg claimed that the 'deductive system' constructed
in this way was 'a distorted version of Durkheim's theory of
division of labor'.(41) In fact it is a set of statements with
indefinite content that apparently resemble some of Durkheim's
theses. Durkheim's theses occur in a definite context, which
consists, in the first place, of Durkheim's reflections, and in
the second place of the broader context in which those reflections must be considered, that is, the sociological discussions
of that period, the style of thinking and the rules of every-day
language of that period, etc. Zetterberg, when proceeding to
construct his 'deductive system', programmatically dismissed
those contextual meanings of Durkheim's theses. Did he impart
them in a different sense? The answer depends on whether we
take his undertaking, as he sees it, seriously or not.
If we assume that the above set of statements is to be interpreted in accordance with the rules of construction of 'axiomatic theories', as earlier formulated by Zetterberg, we face
the necessity of establishing the meanings of the statements in
that set on the basis of the seven 'basic' and three 'derived'
terms. The system is so primitive that no interpretation of it
is possible. Note, however, that for Zetterberg the point of
departure was the set of ready-made propositions, and the

whole operation consisted in manipulating that set so that some
of the propositions could be reduced to the remaining ones.
To do so he arbitrarily assumed that some of the terms occurring in the propositions from that set (later called 'the derived
concepts') would be deduced from the remaining ones (which
thus became 'the basic concepts'). Hence the procedure presented by Zetterberg is an operation on a set of propositions
whose meanings are, in some way, given independently of
'the axiomatic system'. In this case they are given by current
linguistic usage which assigns a more or less definite content
to such terms as behavior, group, solidarity, etc. Thanks to
this, both Zetterberg and the reader who goes through the list
of propositions have the feeling that they know what it is all
about. Moreover, the 'axiomatic system' constructed from that
material inherits the same property, i.e., it gives the reader
the impression of a set of statements having definite sense.
Furthermore, the operation of transforming the original set into
a 'deductive system' is made possible thanks to the fact that
the terms which occur in that set of statements have a rich
semantic penumbra in everyday language. This enables one to
manipulate these terms almost without any restrictions, without
evoking in the reader (or in oneself) a feeling of strain. Uniformity can be defined in hundreds of ways without leaving
the area limited by the current feel of the language. The flexi-
bility of these terms also gives the illusion that the adopted
propositions are true, even in the case of such propositions
as 'The greater the number of members, the more uniformity'
(to tell the truth, it is difficult to say why it should be so, but
perhaps in a sense it is so).

   This example shows in what the danger of using the procedure
suggested by Zetterberg consists. It is a procedure which
makes it possible to transform otherwise sensible propositions
into a pseudo-deductive configuration of empty statements.
Naive ideas about the nature of deductive systems function
here as a 'methodological' screen for arbitrary manipulations.
Zetterberg proceeds as if he did not realize that the meaning
of a statement depends upon its context, and that the occurrence in two statements of one and the same term does not
necessarily prove that the meaning of that term is the same in
both cases. Generally speaking, the practical sense of Zetterberg's proposal boils down to treating statements as building
blocks, of which various structures, including the 'perfect'
one, i.e., the pyramid, can be built. It was a big step back-
wards relative to the methodological awareness of the earlier
generations of sociologists.

Zetterberg's chief influence has been on American sociology.
As Nicolas Mullins puts it,

   the accepted philosophical basis for theory in standard
   American sociology was stated in Hans L. Zetterberg's

Theory and Verification in Sociology, first published in
1953. This very successful little book became the basis for
many graduate courses and exams on theory.(42)

This fact is, it seems, to be interpreted as follows. The
1950s were a period when American sociology had already been
divided into three parts: the methodology of social research,
empirical sociology, and theoretical sociology, or, in brief,
method, research, and theory. Methodology was strongly influ-
encing research, but intellectual connections between theory,
which was then universally identified with Parsons's Grand
Theory, and the remaining two elements were practically nonexistent. Merton's much publicized conception of middle-range
theories was an attempt to cope with that split, an attempt to
bring theory and research closer to one another.(43) The idea
itself was far from being clear and precise: it was not very
clear to what the qualifier 'middle' applied, to the generality
of the theory or to the size of the social groups which it covers,
or to the degree of empirical substantiation. Merton's intention,
however, was clear: his point was to introduce 'intermediate'
research activity between Parsons' super-abstract speculations
and what Mills would later term 'abstracted empiricism'. In
other words, Merton meant activity that would be research-
oriented but open to theory.
    The American sociological milieu received that proposal very
well, for it provided a formula that would mitigate the dissonance between respect for theoretical work and the tendency
to identify sociology with empirical research of a certain kind.
The proposal, however, remained in the sphere of well-accepted
programmatic slogans, for it did not offer any practical recommendations how one should engage in theoretical activity while
accepting the empiricist heritage in toto. Moreover, further
implications of Merton's idea were destructive for the prevailing
tendencies as they introduced theory as an equal partner of
empirical research, whereas 'shallow empiricism', as it was
labelled by some, was based on the assumption that theoretical
thought was derivative with respect to empirical research: a
theory was supposed to emerge in the course of time from the
findings accumulated by empirical research. Now the success
of Zetterberg's idea was closely tied in with the fact that it
answered the question how a theory can emerge from the
achievements of empirical research. It was thus a conception
which solved the problem of the division of sociology into
'research' and 'theory' under conditions fixed by the assumptions that underline 'research'. An additional advantage was
that solution's methodological sophistication (if we assess that
in accordance with the views of the average sociologist) and
its connections with 'true' science, represented by logic and
the methodology of science.
    Be that as it may, the idea of axiomatic sociological theories
in Zetterberg's interpretation came to be included in the durable

attainments of sociology; it can be found (usually together with the above examples that refer to Durkheim) in the most prestigious handbooks of the methodology of sociological research. Zetterberg in a sense initiated the new specialization in methodology, namely one concerned with 'studies' on 'the strategy of theory construction' in sociology. Of course, later works on the subject are on a much higher level than 'On Theory and Verification in Sociology', but they continue the tradition of that work in two very essential respects. First, they analyze properties and the arising of non-existing entities. As the author of one of the best publications of that kind writes, 'I often found myself introducing notions that are unrelated to actual practice, even to the point of describing types of theories that cannot be illustrated by reference to the literature.'(44) Second, the 'deductive system', usually interpreted in accordance with Zetterberg's approach, remains the paragon of sociological theory. Even works concerned with matters that are remote from the problems discussed by Zetterberg stress the idea of a 'deductive system' as the principal one. An example is provided by the publications by Hubert M. Blalock and the authors who are intellectually close to him, concerned with certain techniques of data analysis (causal models). Blalock's book discusses problems which mathematicians would perhaps treat as applied mathematics, but whose links to sociological problems are at best an open issue. Blalock, however, considers his problems to be those of constructing sociological theories; like Zetterberg and Homans he assumes that a scientific theory must be in the form of a set of deductively ordered theorems about relations among variables, that is – and at this point Blalock's interpretation intervenes – in the form of a mathematical model that links a number of variables.

The fortunes of Zetterberg's idea are symptomatic. Within less than twenty years the idea drawn from C. West Churchman's handbook, an idea that seemingly disregarded the realities of sociology, became an accepted and important element of the intellectual training of modern sociologists, an element that was continually expanded and transmitted in the process of teaching. Zetterberg still realized that he was contributing something both new and controversial to sociology. For Homans the 'deductive system' in sociology was already a reality, still disputable but nevertheless a reality. Blalock writes about deductive systems in sociology as about the daily bread of that discipline. For Zetterberg the deductive system was the best form of theory, for Blalock it is already the only scientific form of theory; deviations from the deductive model are interpreted by him as forms which are imperfect and basically transient: 'the kinds of verbal theories that serve as first approximations to deductive theories are often far too simple and unclear to stand as adequate formulations; mathematical models should eventually replace or supplement such verbal theories.'(45)

What is probably most essential here is the fact that an idea once introduced lives its own life, develops into schools and subdisciplines, but does not evolve. When describing a deductive system Blalock does not refer to contemporary philosophical works, but to Zetterberg and Homans, thus repeating the same naiveté and mistakes (and, of course, the same example allegedly drawn from Durkheim). In other words, the idea of deductive theories perseveres in sociology in the same form which it had in a book that was inferior - let us say - to average philosophical publications on the subject.(46)

For all that, the influence of these metatheoretical ideas upon sociology must be assessed as fairly modest. They certainly did contribute to the formation of the methodological superego of sociologists, but the possibilities of putting them into practice proved almost nil, if we disregard trials which had the nature and status of academic experiments.(47) This is, of course, not to say that those ideas did not influence sociological research, but their influence was indirect, through the shaping of criteria of scholarly achievements, and not through undertakings that had the construction of 'an axiomatic theory' in view.

On the whole we must admit that if we were to measure the impact of the doctrine of logical positivism by examining sections of handbooks on the 'methodology of social research', then in the section on 'theory construction' we would have to record adverse effects only, namely destruction of the received mode of theorizing and its methodological foundations. In general, it can be said of logical positivism that, being a doctrine which emerged from a striving to purify science, it was set more on eliminating thinking which was 'alien to true science' than on propagating positive patterns. Nevertheless, its purifying activity of necessity led to some positive proposals: the campaign against 'pseudo-explanations' resulted in working out the model of explanation, the 'purification' of the language of science implied the explanation of the properties of a 'pure' language, etc.

Some of these proposals emerged as it were naturally from the principal assumptions of the doctrine, leading in the course of time to the development of important specializations within the philosophy of science. This cannot, however, be said about the logical-positivist idea of scientific theory. This is quite understandable: logical positivism is basically an anti-theoretical orientation, and there is no need for, or possibility of, studying theories, and particularly their structures, within the framework of that doctrine. As a matter of fact, the term 'theory' should be struck off the list of 'decent' scientific terms. This has not been done because the term was too deeply rooted in physics and other prestigious fields, and its elimination would be an open challenge to those disciplines. Note, however, that in the history of logical positivism that term always marked problems and additional complications, thus forcing the proponents of the doctrine to liberalize their standpoint apparently or in fact (the case of theoretical terms is an example). The conception of theory emerged in the history of logical positivism as a consequence of the complication of the original picture of science.

That process quickly met the natural boundaries delimited by the assumptions of the doctrine; 'the theoretician's dilemma', to use the title of Hempel's well-known work, was that he, the theorist, was faced with the necessity of choosing between his own methods and techniques and what he could afford under the most liberal possible interpretation of the doctrinal assumptions. Hence the logical-positivist proposals concerning theories boiled down to rearguard action intended to save the doctrine in the face of the less and less favorable intellectual situation. It is not surprising, therefore, that they were exceptionally poor. In any case, they were not suggestions that would give a willing 'working scientist' a positive research program. It is true that sociologists have deducted a whole 'strategy of theory construction' from those suggestions, but, first, they have managed to do so owing to a superficial knowledge of those suggestions and a lenient approach to the rules of inference, and second, that strategy - as has been mentioned earlier - has so far yielded nothing remarkable.

Be that as it may, the issue of 'theory construction' is a good opportunity for tracing the process of permeation of sociology by the logical-positivist philosophy of science, but gives no opportunity for showing practical applications of those ideas. We shall therefore pass to another, but related, field, namely that of the conceptual apparatus, and in particular problems connected with the requirement of the empirical meaningfulness of (extra-logical) terms used in science.

As has been mentioned on several occasions, logical positivism evolved toward an increasing liberalization of originally very rigorous methodological ideas and requirements.(48) The most striking modifications were those concerned with scientific terms. Something of basic importance for the doctrine as a whole, was at stake, the more so as the search for an adequate criterion of verifiability in terms of logical connections with observational statements had failed despite successive modifications of the criterion suggested, and further attempts aimed at defining the properties of the language of science: verifiability was to be the property of statements formulated in language that complied with definite conditions. Those conditions covered: (i) the syntactic rules of the language (ii) the terminology used in that language. The conditions imposed upon the terminology, and, to put it more precisely, upon the extra-logical terms occurring in the language, were of decisive importance. It was these conditions which underwent essential modifications up to the point where any further evolution (liberalization of the conditions) would have threatened the basic assumptions of the doctrine. To simplify the discussion let us list the successive stages.

1 At first it seemed that the terminology of empiricist language should include (a) logical terms, (b) observational predicates that would form the empirical basis of the language, and (c) expressions defined in terms of (a) and (b).
2 It soon turned out that the terminology that would meet those conditions would be too limited, and that, for instance, it could not comprise the language of physics. This was the conclusion drawn from the analysis of dispositional terms; their definitions encountered formal problems, successfully solved by Carnap, but at the cost of a modification of the conditions originally imposed upon the terminology. The modification consisted in admitting terms introduced with the aid of reduction sentences which, while not being definitions in the strict sense of the word, have the nature of partial definitions. In this way 'empirical constructs', i.e., terms which have only a partial specification of meaning based on observational predicates, were admitted.
3 The third stage brought an essential change in the interpretation of the relation between observational predicates and the other extra-logical terms. This was due to the necessity of taking 'theoretical constructs' into account, which can be introduced neither by full nor by partial definitions, their meaning being established by the assumptions of that theoretical system within which they function. Their relationship with observational predicates is difficult to explain. It is, in any case, an indirect relationship and such that it is rather the theoretical constructs which specify the meanings of the observational predicates than conversely.(49)

What was the process of making the language of sociology 'empirical', as seen in the light of the evolution outline above? The beginnings of that process were connected with the discussion of operationism, which in sociology started almost immediately after the publication, in 1927, of the 'Logic of Modern Physics' by P.W. Bridgman and continued for nearly thirty years to wither slowly and unproductively, if the productiveness of a discussion be measured by its conclusiveness. Today it is self-evident that the controversy over operationism was undecidable under the then existing conditions and had to end as it did in a sterile exchange of arguments. On the one hand, Bridgman's conception was untenable regardless of the tricks of interpretation used, because it led to conclusions that were either absurd (if rigorously interpreted) or glaringly trivial (if, as occurred later, more liberally interpreted). On the other hand, operationism fitted perfectly well with the general tendency then prevailing in sociology (and in psychology as well), and fighting against it meant swimming against the stream. In the whole dispute, arguments carried more weight than reasons, the point was to win in the discussion and not to solve the problem, as is usually the case when essential issues connected with the raison d'être of some orientation is at stake.

In this case the issues involved were of vital importance for the emerging 'scientific' orientation. It was, of course, in a sense a coincidence that the controversy was over operationism: it could just as well have developed over any other conception that was convenient from the point of view of 'scientism'. In fact, the discussion of behaviorism was taking place at the same time, among the same participants, on the basis of the same arguments, and with the same results (i.e., without any result). The two controversies were closely connected. Had Bridgman not raised the idea of operationism, all attention would have been focused on behaviorism, or physicalism in Carnap's sense, or any other similar conception would have taken the place of operationism. Bridgman, who probably thought he was writing for physicists, offered the sociological and psychological milieu an excellent occasion for articulating the debatable issues which abounded in these two disciplines. The occasion was made even better by the fact that the operationist standpoint had never been formulated by the operationists themselves in an orderly manner, which made it possible to pass from one interpretation to another and thus to raise a vast range of issues. The discussion which was seemingly concerned only with a technical issue (how to define concepts) was in fact a clash between two theoretical and methodological orientations which vied for the first place in sociology. Therefore the attacking party at least was not interested in restricting the subject matter of the dispute by making the concept of operational definitions precise. Lundberg protested against the 'manoeuvres' of the opponents who demanded an unambiguous description of operational definitions (e.g., through an operational definition of operational definitions), because, as he claimed, such demands led only to production of 'words about words, about words, etc.', which was a task fit for the opponents of operationism, but not for operationists themselves.(50)

Lundberg was right in defining the parties to the controversy as positivists and antipositivists. Operationalism, like behaviourism, developed outside the philosophy of logical positivism, but, again, like behaviourism it was able to successfully articulate some logical-positivist ideas. While behaviourism can be interpreted as an attempt to put the requirements of physicalism into effect, operationism can be viewed as a technical version of the requirement for the empirical interpretation of scientific terms. But that version introduced certain unnecessary (from the logical-positivist standpoint) restrictions and, if only for this reason, could not become part of logical-positivist doctrine. In any case operationism has remained outside the system proper of logical positivism as a slightly embarrassing, boisterously radical ally.(51)

There is no reason to further concern ourself with an exposition of operationist assumptions or to relate the discussion of operational definitions in sociology.(52) Little is left of the

original operationism in sociology, and the whole issue would
be an insignificant episode in the history of social thought were
it not for the fact that the dispute over Bridgman's operationism initiated 'operationism' in sociology. Bridgman's words: 'I
feel that I have created a Frankenstein, which has certainly
got away from me'(53) have the value of a profound truth. Like
most methodological ideas born outside sociology, operationism,
when transplanted to sociology, gradually started living its
own life and evolved in a direction determined, on the one hand,
by the gradually vanishing familiarity with the source material,
and on the other hand, the needs of the sociologist's research
practice. This came to shape an attitude which, while in general
preserving the characteristic tendency of operationalism, in
detail moved well along toward common-sense dilettantism. To
present that attitude we again have to refer to Zetterberg's
'On Theory and Verification in Sociology', for it was there that
it was first formulated in the form which has survived to this
day with very minor changes and is now the foundation of the
standard research procedure in 'empirical' sociology.

Zetterberg opposed operationism understood as the standpoint which accepted operational definitions as the only ones
valid in science. Science, he claimed, needs nominal definitions
as well, for without them one cannot engage in theoretical
considerations. Moreover, 'it is plain that only those operational
definitions that have a counterpart in a nominal definition are
worthwhile.'(54) On the other hand, he says, the process of
verification of (theoretical) hypotheses would be impossible
without operational definitions. Thus, while an operational
definition without its counterpart in a nominal definition is
worthless, a nominal definition which is not accompanied by
a corresponding operational definition provides no opportunity
for empirically verifying the given hypothesis. 'From the above
it appears evident that one worthwhile aspect of the operation-
alist movement is an appeal to keep nominal definitions and
operational definitions in closest contact.'(55)

But what is that contact to consist of? The point is that the
operational definitions should translate what is contained in
the nominal definition into a language that is 'more acceptable
for research.'

> We may, for example, select the number of occupations to
> stand for the degree of division of labor. And we may
> select the proportion of laws requiring the death penalty,
> deportation and long prison terms (but not fines) to stand
> for the degree of rejection of deviates from society of
> norms.(56)

Zetterberg in a way fundamentally moderates the operationist
standpoint. In the peak period of the discussion of operationism
in sociology, the terms introduced by the other means than
operational definitions were treated as 'noises, which have an

xpressive function comparable to exclamations of joy or sad-
ess, laughter, or lyric poetry, but which have no objective
epresentative function at all.'(57) Zetterberg not only rejects
hat approach, but treats an operational definition not as an
peration intended to impart meaning to a given term, but as
n operation in the sphere of research pragmatics: an opera-
ional definition (or perhaps rather the operational defining
f the term involved) is just a step in the process of verifying
given hypothesis. It is an indispensable measure, that is
rue: if one wants to verify a hypothesis, one has to formulate
t in a language 'more acceptable for research'. This is not to
ay, however, that without such an operation the hypothesis
eing subject to verification is meaningless. For a sociologist,
perationalization is just a *practical* necessity.

Ascribing a purely pragmatic function to the operational
efinition weakened the doctrinal importance of operationism
nd reduced its connections with logical positivism to technical
atters. At the same time, however, it created the conditions
or the transformation of the still elitist operationist movement
nto a popular one. The average sociologist can grasp technical
ecessity much more easily than epistemological ones, and the
rguments showing why the researcher's first step must be 'to
ranslate hypotheses into the language of research operations'
ppeal to him much more than arguing about the meaninglessness
f everything with which sociology has been concerned up to
ow.

Neither Zetterberg nor any of his followers has formulated
recisely which terms are 'more acceptable for research', in
ther words, into which language a 'nominal definition' is to be
ranslated if an 'operational definition' is to result. In one place
etterberg writes 'operational we call the definitions that refer
o descriptions of measurements or enumerations', but this is a
ypical example of a definition that misinforms about its author's
ntentions. It seems that Zetterberg formulated it under the
nfluence of the tradition which (verbally) linked operational
efinitions with measurements. In any case, both Zetterberg's
xamples (the statement 'I like it here in X-town' as an opera-
ional definition of 'work satisfaction') and the later practice
hows that the language of operational definitions is the language
f standard empirical research, that is, mainly the language of
ociological questionnaires, further the language of statistical
ata, clinical observations, psycho-social experiments, etc.
perational defining of terms (i.e., 'operationalization' of
roblems and hypotheses, i.e., their translation into 'the lan-
uage of research operations') thus consists in reducing those
erms to an observational language determined by the research
ractice of a typical empirical sociologist, that is, to a language
hich a typical empirical sociologist treats as the language of
acts, the language of empirical data, the language of obser-
ations. This means that for an empirical sociologist such a
eduction was not only a reduction to those categories which he

used in his research practice; for him it had a profound
methodological meaning as a reduction to the language which
he identified with the language of empirical research in
general.

But the average sociologist saw such a reduction as an
attractive prospect for the realization of his theoretical ambi-
tions without going beyond standard empirical research. Moreover, the new version of operationism gave every researcher
an opportunity for confining himself to those categories with
which he felt most at home. Thus an expert in surveys could
'operationalize' practically all the things that he was concerned
with in the language of sociological questionnaires; a person
accustomed to the language of statistical data could translate
his problems into the language of those data; and a social
psychologist – regardless of his subject matter – did not have
to go beyond the language of his quasi-experiments. For the
requirement was: every theoretical problem (hypothesis, term,
etc.) should be translated into the language of research
operations. The range of those operations (in other words, the
answer to the question what those research operations are)
was given by the research practice of standard empirical sociology, but a given sociologist's choice within that range of
operations was his own matter. It is true that an additional
requirement was introduced, namely the requirement that such
a translation be a valid one (in Zetterberg's terminology it was
formulated as the requirement for the validity of an operational
definition), but the requirement could always be so interpreted
that the validity of the translation would be assumed within the
framework of the empirical research of a given type. Thus if
the choice (determined by the researcher's experience) was
to resort to questionnaire studies, i.e., if he decided to carry
out the 'operationalization' on the basis of the opportunities
offered by the questionnaire, then the problem of the validity
of the translation was: is a given questionnaire a good one, or
must it be replaced by another one (i.e., in practice, is a give
questionnaire to be corrected or emended?). This seems the
best explanation of the mania for questionnaires in sociology
over the last thirty years. It is true that the questionnaire is
an exceptionally convenient instrument, but for a convenient
instrument to become the dominant one there must be a methodo
logical doctrine which sanctions it. To put it in more concrete
terms, if the majority of sociologists almost always associate
the operationalization of a problem (and that regardless of
what the problem is) with the construction of a questionnaire,
this happens because the conception of operationalization
permits it to.

But let us return for a while to the issue of the validity of
operational definitions. Zetterberg writes that 'all operational
definitions we have to judge with regard to reliability and
validity' (we can disregard reliability here, because from the
point of view of the problems we are concerned with, it is

a neutral issue of a technical nature), and defines validity as the extent to which an operational definition corresponds to a nominal definition.'(58) This means that, theoretically, an operational definition transfers the meaning of the term that occurs in theoretical hypotheses onto another, empirical, level, and hence it is not a definition in the strict sense of the word, i.e., an expression which establishes the meaning of the term. If this is so, then the assessment of the validity of an operational definition is in fact of primary importance and, let it be added at once, a very intriguing one: the whole program of operationalization' would have to start from the question how such a transfer of meanings of the terms used in a given theory is possible, all the more since Zetterberg in principle requires a full explication of the concepts used in a given theory in terms of 'research operations' and treats deviations from perfect validity' as errors. Unfortunately, neither Zetterberg nor his followers have answered that question, and there are no reasons to believe that they have even noticed the problem.

This accounts for the fact that their comments on the validity of operational definitions are rather meaningless generalities. Specifically they practically ignore the issue of how one can make sure, and if so, to what extent, that a given operational definition is valid. All we find on that subject in Zetterberg's book is what can be inferred from the following two sentences:

> The major difference between internal validity [i.e., the validity of operational definitions] and external validity [i.e., the truth of hypotheses] is that the former expresses a 'logical' relationship while the latter expresses an empirical relationship. Internal validity, in other words can be appreciated without empirical studies, while the determination of external validity is a test of a hypothesis.(59)

This explicitly refers to the then fairly popular conception of the so-called logical method of evaluation of the validity of a scale,(60) which method is based on the idea that 'the items being what they are, the nature of the continuum cannot be other than what it is stated to be.'(61)

As follows from the context, this idea should be interpreted thus: a careful consideration of the content of the questions (quite understandably, the validity of a scale is usually analy-zed with reference to questionnaire studies) allows one to see whether they have been pertinently chosen and formulated – pertinently from the point of view of the problem in question. It is thus a method which everyone uses – although without being conscious of the fact that he is applying the logical method of testing the validity of a scale – for instance, in an everyday conversation. Good and Hatt correctly state that the logical validation (that is the evaluation of validity) is almost always used. Since other known ways of evaluating the validity of a scale are particular varieties of the 'logical method', the

problem in practice is solved by a more or less subtle analysis
of the various items in the questionnaire. As I have shown
elsewhere, (62) it is one of those numerous points where to
'scientific', 'anti-intuitionistic' and 'rigorous' sociology is added
intuition, the ability to put oneself in the respondent's place,
etc.; since this addition is not consciously perceived, there
is virtually no possibility of a conscious control of the process.

Thus, in practice, the requirement for the validity of operational definitions does not play any major role as a barrier
against arbitrary translations. Where the questionnaires used
are for studying the everyday experience of people in the same
socio-cultural sphere as the researcher himself, checking the
validity of the questions relative to the problems studied is
fairly effective. But if researchers 'operationalize' more complex
theoretical concepts, especially those referring to social macro-
structures, the relationship between 'an operational definition'
and the original concept is usually merely stated (such a statement not being usually given the status of a meaning postulate).

Since the first edition of 'On Theory and Verification in
Sociology' appeared (though of course not only due to that
book, which articulated rather than initiated a certain trend),
'operational defining' of the facts studied has become the basic
operation in the standard research procedure in sociology.
Zetterberg split that procedure into eleven steps:

(1) the creation of operational definitions corresponding to
the nominal definitions of the hypothesis; (2) the evaluation of the validity of these operational definitions; (3) the
evaluation of the reliability of these operational definitions;
(4) the translation of the original hypothesis into a working
hypothesis; (5) the selection of the population on which the
test of the working hypothesis shall be performed; (6) the
evaluation of the scope of the population; (7) the drawing
of a sample from this population; (8) the evaluation of the
representativeness of the sample; (9) the test of the working hypothesis; (10) the evaluation of the outcome of the
test of the working hypothesis; and (11) the acceptance or
rejection of the original hypothesis. (63)

That schema reappears, more or less exactly, in successive
generations of handbooks of 'the methodology of sociological
research'; the modifications consisting in different arrangements
of the procedure described by Zetterberg's eleven steps rather
than in any essential changes. Note that the above procedure
can also be presented as a logical sequence of research operations of three kinds:

(i) 'operationalization' of the problems to be investigated,
i.e., translation of those problems into 'the language of
research operations', which implies replacement of the 'original
hypotheses' by 'working hypotheses';

(ii) gathering and processing of the data determined by (i) (which includes such operations as the choice of the sample); (iii) the operation converse to 'operationalization', i.e., translation of the outcome of the study back into the language of the 'original hypotheses'.

The operations mentioned under (ii) are the traditional domain of sociography, i.e., collection and processing of information on the basis of simple and current social knowledge. The methodology represented by Zetterberg among others introduces operations (i) and (ii) as bridges between sociography and theoretical thought. But the essence of those operations boils down to an attempt to ennoble sociography by making it apparently dependent upon theoretical thinking. By offering empirical researchers a 'methodological basis' for an arbitrary translation of theoretical content into a language of research operations that methodology gave the new sociography the appearances and rank of creative work of essential theoretical importance. On the other hand, by formulating the absolute requirement of 'operationalization' that methodology gave the new kind of sociography a monopoly position in sociology. The success of that methodology meant a withering of theoretical thinking, and the elimination from sociology of content that is not translatable into some set of standard empirical data.

To sum up, we have to say that operationism in Zetterberg's version has become the methodological foundation of the process which gradually turned sociology into a discipline that mass produces theoretically trivial, if not incoherent (and hence useless), data. The process poses a serious threat, all the more since it takes place behind a façade of pseudo-theory which makes it hard to expose its true nature. This façade accounts for the fact that common sense, which in traditional sociography protected researchers from accumulating quite useless information, has lost that control function in the new sociography: behind the screen of operationalization of some theoretical problem or other, research projects are taken up, which are completely at variance with both theoretical thought and common sense. The fact is too well known (although it is usually given a different explanation, namely that it results from the wrong implementation of a certain idea in sociological research, and is not the natural result of the idea itself) to both sociologists and outside observers to require exemplification.

As has been mentioned, Zetterberg did not link operational definitions to the issue of the empirical meaningfulness of terms; in his approach, theoretical terms require operational definitions because empirical research is, by the very nature of things, conducted on the operational level, and hence we have to pass from the theory level to the 'operational' level, later to return to the theory level again. But it is possible, while accepting that opinion, to go a little further and to try to support Zetterberg's view with epistemological arguments.

Here is an example of such an argumentation.

> A concept is an abstraction from observed events or, as
> McClelland ... puts it, 'a shorthand representation of a
> variety of facts. Its purpose is to simplify thinking by
> subsuming a number of events under one general heading.'
> Some concepts are quite close to the objects or facts they
> represent ... . Other concepts, however, cannot be so
> easily related to the phenomena they are intended to repre-
> sent; *attitude, learning, role, motivation,* are of this sort.
> They are influences, at a higher level of abstraction from
> concrete events, and their meaning cannot easily be conveyed
> by pointing to specific objects, individuals, or events.
> Sometimes these higher-level abstractions are referred to
> as *constructs,* since they are *constructed* from concepts
> at a lower level of abstraction.
>   The greater the distance between one's concepts, or con-
> structs, and the empirical facts to which they are intended
> to refer, the greater the possibility of their being mis-
> understood or carelessly used, and the greater the care
> that must be given to defining them. They must be defined
> both in abstract terms, giving the general meaning they
> are intended to convey, and in terms of operations by
> which they will be represented in the particular study. The
> former type of definition is necessary in order to link the
> study with the body of knowledge using similar concepts or
> constructs. The latter is an essential step in carrying out
> any research, since data must be collected in terms of
> observable facts.(64)

This quotation comes from a handbook of something which
(following the suggestion of Abraham Kaplan) we would call the
technology of sociological research, and hence a handbook
concerned with technical issues and seemingly quite neutral
on controversial issues of epistemological nature. This handbook is one of the most widely used texts on methodology in
sociology, and can well be treated as an interpretation of the
views of the average sociologist in the 1960s and - with some
reservations - in the 1970s. The concurrence between the
opinions of its authors and Zetterberg's view is obvious, but two
characteristic differences must be noted. In Zetterberg's interpretation, a nominal definition is part of a 'deductive system'
(or at least some theory), whereas in the interpretation of
Selltiz and her co-authors a definition 'in abstract terms' (and
hence the counterpart of a nominal definition) like an operational definition (which they prefer to call a working definition)
is formulated ad hoc, in connection with a given research project, and hence is rather part of that project than of any theoretical system. Second, in Zetterberg an operational definition
is a special case of translation of a nominal definition, while in
Selltiz et al. a nominal definition explains the meaning of a

given term in abstract terms, and a working definition, in terms of 'research operations' or 'observable facts'. This can be reformulated thus, in agreement with the context: a nominal definition explains the 'abstract' meaning of a given term, while a working (operational) definition, its operational (concrete, empirical) meaning. Nominal definitions function as bridges between empirical research and an unspecified 'body of knowledge using similar concepts or constructs'; the role of operational definitions, if we are to infer from the first part of the quotation, is much more essential since they specify those empirical elements from which a given abstract concept is constructed, or at least those elements which will 'represent' that concept in a particular study. In the light of the theory of concepts, outlined in the above quotation, this approach assigns to the operational definitions the status of a sentence that establishes the empirical meaning of terms.

It must be admitted that the foregoing interpretation is largely based on conjecture and the feel of the context: it is not easy to reconstruct the intentions of the authors by sticking closely to a text that is full of gaps in reasoning, contradictions etc. But I think there is no doubt that we are dealing here with the idea of a double meaning of 'abstract' terms in sociology: the abstract meaning, which should perhaps be called theoretical meaning, and the operational meaning, which could, as seems to accord with the authors' intentions, be called empirical meaning. That idea has found broad application in practice by giving to the requirement of operationalization epistemological sanction, which in some respects is stronger than pragmatic sanction. After all, the epistemological sense of operationalization, i.e., the treatment of operationalization as a procedure for imparting empirical meaning to terms, has been present in sociology since the discussion of operational definitions. It was dismissed in Zetterberg's works, but in broad circles of working sociologists the operational definition, however it was understood, was irreversibly associated with a more or less clearly felt idea of the empirical meaning of terms (as something which must be imparted to terms). As the influence of logical positivism was spreading, the concern for imparting empirical meaning to sociological terminology was taking on the form of more and more consciously held methodological conceptions, to be ultimately fixed in the form represented, for instance, by the following statement:

When a student of social phenomena proceeds to verify a hypothesis, *he faces the task* of translating the usually rather general concepts, included in that hypothesis, such as attitude, orientation, social structure, and socioeconomic status, into the language of those phenomena and variables which he will observe.

At this moment he enters the sphere of problems which philosophers of science and methodologists call correspond-

ence rules, epistemic correlations, operational definitions,
etc., which pertain to the relation between the language
of theory and the language of empirical research. Terminological differences seem to be primarily verbal, the purpose
of the various ideas that use those concepts being the same:
to guarantee the empirical sense of scientific terms.(65)

Thus, aside from pragmatic justification, the requirement
that 'concepts', 'problems', 'hypotheses', etc., be translated
into the language of 'research operations', 'observable facts',
'empirical data', etc., that is, that everything which is formulated in a different language be translated into the current
language of standard empirical research, this requirement had
strong epistemological sanction as well. From the point of view
of our present problems, we may treat the concept of operationalization as the idea of solving the problem of the language of
sociology in accordance with the requirements of logical positivism.

The requirements of logical positivism from which period?
Now the average sociologist still takes the word 'translation'
literally, so that in the process of operationalization of a
research problem the terms used in the 'theory' are *replaced*
by terms used in 'research operations'. As a result, the original problem completely vanishes from the researcher's field
of vision; in exchange science is presented with an ad hoc
formulated question, about relationship between 'observable
variables', such as answers to specified questions in the
questionnaire. In other words, when it comes to his conceptual
apparatus, the average sociologist to this day represents the
standpoint that corresponds to the views of philosophers of
science in the earliest period, when empiricist requirements
concerning the language of science were not yet complicated by
the amendments introduced by Carnap in 'Testability and Meaning'. This is, by the way, quite understandable. The simplest,
and most radical, conception of empiricist language has all the
advantages of common sense and very easily lends itself to
popularization. The chronology of events was also significant.
The ideas of the philosophy of science usually affect sociology
with a time lag of about twenty years. The logical-positivist
breakthrough in sociology occurred in the late 1940s and the
early 1950s. The ideas which were being transferred in that
period and which came to shape sociologists' current methodological consciousness - which was soon to become objectified
and fixed in standard texts serving as handbooks and thus
transmitted from generation to generation as the scientific
foundations of sociology - thus originated in the early 1930s.
The claim that the main trend in present-day sociology is
based on the logical-positivist ideals of science of nearly fifty
years ago is a simplification, but it is not falsehood. Every
philosopher of science, be he ever so faithful to logical-positivist tradition, can realize how anachronistic that foundation is.

The history of sociology over the last twenty years - at least
when it comes to the main empiricist trend in that discipline,
with which we are here concerned - in a sense consists of
attempts to modernize the methodological conceptions which date
from the Vienna period. These attempts, of course, follow a
course which accords with the evolution of logical positivism,
and are undertaken with necessary time lag. Thus the mid-
1960s witnessed the idea of an empiricist language of sociology
which was an attempt to overcome the approach that marked
the philosophy of science at the time when die physicalische
Sprache was being discussed. This idea cannot easily be localized in the history of logical positivism as it is marked by a
confusion of the various issues and opinions, not to speak about
simple inexactitudes in the discussions: such categories as
theoretical term, correspondence rules, observation term, etc.,
were used quite freely, and, for instance, the problems of
partial definitions were incorrectly expounded both from the
substantive and formal point of view. At any rate, these conceptions remain within the boundaries marked, on the one
hand, by the emergence of the problem of the partial definability of some scientific terms, and on the other hand, by the
beginnings of the discussion of the relation between the language of theory and observational language. These conceptions
can, with some reservations, be defined as an attempt to
introduce the concept of a theoretical language that is not
reducible to an observational language. Characteristically,
they usually refer directly to philosophers of science, mainly
Hempel and Carnap, less frequently to Popper, but - with a
few exceptions - leave the works on the subject by Abraham
Kaplan out of account, although his works function in principle
as part of sociological literature. (66)
  And here is an abbreviated version of one such conception. (67)
  The authors begin with stating that 'philosophers of science
have recently raised some challenging and controversial
methodological issues concerning "theory" [note the quotation
marks] in the social sciences', and that 'according to rigorous
[the favourite and oft-repeated qualifier] evaluative criteria'
provided by the philosophy of science, sociological theories
leave very much to be desired. (68) In the authors' opinion,
this diagnosis can lead to the adoption of one of the following
two procedures:

  (1) purging the realm of sociological inquiry of all that does
  not conform with the standards of rigorous evaluative criteria, or (2) developing a practical program for selecting
  or developing those aspects of current sociological theory
  that *show promise for eventual conformity with these
  criteria.* (69)

Since the first procedure 'is likely to be enshrouded with
numerous *impracticalities*' (author's italics), the authors are

inclined to adopt the program of saving what can be saved in
sociology 'according to rigorous evaluative criteria', or, to
put it more precisely, of indicating the course that would help
raise sociological theory to a higher level.

Dumont and Wilson make a distinction between an explicit
theory, a theory sketch and an implicit theory. They identify
an explicit theory with an interpreted axiomatic system. In
such a theory, they say, theoretical terms have both an epistemic and a constitutive meaning.

> The former signifies that the concepts are connected, either
> directly or indirectly, with observables by rules of cor-
> respondence that have been empirically justified, i.e., via
> these rules, confirmed relationship have been established
> between observable concepts and theoretical concepts [here
> they refer to texts by Carnap and Torgerson]. The latter
> implies that the concept in question enters into a sufficient
> number of relations with other terms in the theoretical laws
> of the postulate network, and contributes to the explanation
> and prediction of observable events.(70)

Sociology, as they correctly claim, has no such theories at
its disposal, and hence 'theoretical concepts in sociology lack
both epistemic and constitutive significance.'(71) In sociology
there are, however, 'scientifically acceptable' theory sketches,
one of them being 'Homans's theory of elementary social behavior'. A theory sketch is the same as an explanation sketch as
understood by Hempel.(72) The concepts which occur in a
theory sketch lack both epistemic and constitutive significance,
but they have epistemic and constitutive connections, because
they are linked to other 'theoretical' (quotation marks by
Dumont and Wilson) concepts of the theory sketch and with
observation terms, even though those connections are weaker
than in the case of an explicit theory. Thus,

> although concepts embedded in theory sketches lack epis-
> temic and constitutive significance, they are part of test-
> able propositions, and can be evaluated according to the
> nature of their constitutive connections with other theo-
> retical terms and their epistemic connections with *observ-
> able terms*. In each instance they may be accorded 'poten-
> tial significance'.(73)

In the authors' opinion, most concepts used in sociology are
not even included in theory sketches. They define such concepts as 'isolated abstract concepts'. Are they worth anything?
That depends, they say, on whether they can be explicated
so that they at least meet the conditions satisfied by the concepts which occur in a theory sketch. Some isolated abstract
concepts find support in the 'implicit "theory" that awaits
formal discovery before its forms, content and function can

be made explicit.'(74) Such concepts aspire to the status which the terms that occur in a theory sketch possess. Explication – understood after Hempel as a unique reinterpretation of the meaning of a given term – of an isolated abstract concept, results in the articulation of a given implicit theory, i.e., in its transformation into a theory sketch. In other words, 'isolated abstract terms' have scientific value insofar as they are at least potentially concepts of a theory sketch. Of course, 'it would be unrealistic to assume that all *prescientific* (author's italics) candidates for the theoretical language are amenable to such explication.'(75) How then are we to treat those 'isolated abstract concepts' which do not stand the test of explication? Dumont and Wilson do not state that explicitly, but their text offers the unambiguous answer: those concepts stay outside the sphere of science.

There is no need to go into the details of that idea nor to consider its agreement (claimed by its authors) with the recent views of the leading philosophers of science on theoretical terms. We should note, however, that Dumont and Wilson apply their 'rigorous evaluative criteria' in a way which is at variance with their purpose. In the philosophy of science we notice recently a certain (very desirable, though belated) restraint in making statements on the social sciences and other 'less advanced disciplines', and in particular we observe caution in assessing those disciplines on the basis of one's own models. This is, unfortunately, not to say that there has been a revision of the opinions on the nature of the said disciplines: rather, interest in those problems has declined in favour of concentration on internal problems of the philosophy of science. In accordance with that general tendency, theses on the language of theories are formulated as if their scope were limited to formalized axiomatic systems in the empirical sciences, while questions about the language of those theories which are looser in form than formalized axiomatic systems are left unanswered – in particular such problems as how the meaning of extra-logical non-observational terms is established in such theories, whether those terms deserve to be called theoretical, etc. Such reticence is good, at least in the sense that the sociologist who is attached to logical positivism seems to have some leeway in adjusting his philosophical doctrine to the realities of his own discipline. It could accordingly be assumed that the concept of the language of theories, e.g., in the version formulated by Carnap, applies to the social sciences only by analogy, and that its full application to those disciplines would require an appropriate broadening interpretation. Such an interpretation could be that the meaning of theoretical concepts are established by a given theoretical system, whatever its properties, and in particular its structure, might be. In the case of such an approach, Carnap's proposal could be treated as the demonstration of how in certain theoretical systems, namely in the formalized axiomatic systems, the meanings of theoretical terms are determined by a given system, in particular, how such a system links theoretical terms with observational terms. One could accordingly argue that in the theoretical systems of the social sciences, connections between theoretical terms and observational terms (whatever that is to mean) differ from (are much more complex than) the connections described by Carnap, and that therefore such expressions as 'rules of correspondence' when applied to those systems can have only a metaphorical meaning. Note that Abraham Kaplan's arguments go in the same direction:

> For theoretical terms a full definition by observables is even in principle impossible. Carnap speaks of 'rules of correspondence' or 'correspondence postulates' which give what he calls a 'partial interpretation' to theoretical terms. As I shall point out below, what is important is not so much that the interpretation is incomplete as that the observations do not give meaning to the theoretical term but rather mark the occasions for its application. Its meaning derives from the part that it plays in the whole theory in which it is embedded, and from the role of the theory itself ... . 'Castration complex' is meaningless if dissociated from psychoanalytic theory.(76)

It is disputable whether such a far-reaching reinterpretation of the conception of the language of theories can be brought into agreement with the basic assumptions adopted in logical positivism. It remains a fact, however, that the discussion of the language of theories has opened interesting prospects for a further evolution of views on that subject, and in particular for bringing those views closer to the realities of the social sciences. It is a different matter whether such an evolution is possible without infringement of the very foundations of logical positivism, and whether the parties to the discussion would be willing to accept such a trend of the evolution.

All this shows that Dumont and Wilson are even more Catholic than the Pope. To statements made by Carnap and Hempel they add theses which neither of these expounded, at least explicitly, and which no serious philosopher of science in the 1960s would be willing to accept in toto, even if they turned out to follow from the views adopted on the language of theories. They are primarily the following theses:

1 sociology lacks theory (it has pseudo-theories, implicit theories, and, at best, theory sketches);
2 sociology lacks theoretical concepts (some concepts used in that discipline allow one to hope that they can be transformed into theoretical concepts);
3 we have to gradually replace everything that now passes for sociological theories by true theories, i.e., axiomatic systems (with interpretations), and what passes for theoretical concepts, by truly theoretical ones.

Dumont and Wilson's use of the conception of the language of theories is a perfect example of dogmatic treatment of methodological ideas and of deducing from them evaluations and requirements which are as arbitrary as they are absurd. It is not a new style of thinking, but if it is adopted by sociologists and reflected in the most prestigious journals and 'modern' handbooks,(77) we have to conclude that sociology is facing a crisis as a result of the decline of its practitioners' ability to think independently.

Dumont and Wilson's paper belongs to the second generation of methodological publications in sociology which draw ideas directly from the philosophy of science, and that from comparatively recent sources, so that they are believed to be both sophisticated and modern. But let us consider an example of adaptation of the logical-positivist ideas of the language of science made at the really highest level, as compared with other endeavors of that kind.

Andrzej Malewski, when engaged in a controversy with Maria Ossowska over Festinger's concept of cognitive dissonance, once claimed that it was a theoretical concept (in Carnap's sense), and hence neither had to, nor could, meet the requirements formulated by Ossowska.(78) 'Like other theoretical terms the term "cognitive dissonance" does not have any full empirical interpretation in L. Festinger's theory,' but it has a partial interpretation owing to the 'criteria of applicability' which link 'the term "dissonance" with corresponding observational terms'. In Malewski's opinion, the criteria of applicability of the term 'cognitive dissonance' are established in Festinger's theory inter alia by the following formulations:

(a) if a person makes a decision, he is likely to experience a dissonance;
(b) if a person acts under the pressure of a reward or punishment in disagreement with his own beliefs, he is likely to experience a dissonance;
(c) if a person encounters opinions which are incompatible with his own, he is likely to experience a dissonance.(79)

I think Malewski made two mistakes here. First, such formulations as 'decision making' and 'encountering opinions incompatible with one's own opinions' can in no wise be treated as observational terms. We could at most, by straining our interpretation, treat them as dispositional terms and as such include them in the observational language in the broader sense of the word. The second error is more serious. A given expression is a meaning postulate not on the strength of some author's decision, but on the strength of the function it performs in a given theoretical system. Are the statements quoted above postulates in Festinger's system? The question seems to be badly put, since Festinger's theory does not satisfy the formal conditions that would enable us to specify the function of the statements in the system in the way the idea of meaning

postulates assumes. More generally speaking, I do not think
it is legitimate to analyse 'loose' theoretical systems in such
terms as 'meaning postulate', 'rule of correspondence', etc.
This is why Malewski's interpretations must be considered
quite arbitrary.

Five years before that discussion, Malewski made his well-
known interpretation of historical materialism in terms of the
assumptions of logical-positivist methodology.(80) His interpretation was – incorrectly, I think – received as a criticism
of historical materialism. Whatever Malewski's intentions were,
his text was aimed at all classical theoretical systems in the
social sciences. Malewski's procedure could equally well be
applied to the theoretical analyses of Max Weber, Pareto,
Mannheim, Veblen, Simmel, Sorokin, Gunnar Myrdal, Riesman,
de Tocqueville and others. His argumentation aimed to show
that in theories like historical materialism we can separate
parts that meet the requirement of empirical meaningfulness
from those which do not meet that requirement. Thus, for
instance, the empirical meaning of the theory of the class-
based function of the state and the theory of the class-based
nature of law (in Malewski's terminology) can, in his opinion,
be expressed by the following three hypotheses:

1 For every group within society: the richer it is, the more
indispensable it is in the process of production, the more
numerous it is, the better organized it is, and the greater the
armed forces it has at its disposal, the stronger it is.
  This is completed by the following commentary:

> The empirical meaning of the formulation stating that a group
> is stronger than another can be explained by the postulate
> stating that if a group A is stronger than a group B, then
> it can make its aspirations come true even against the
> aspirations of group B.

2 If one group within a given society is stronger (in the sense
indicated above) than any other group in that society and
the binding laws are at variance with its interests, then those
laws are changed or are not enforced in practice.
3 If one group within a given society is stronger than any
other, and if the government acts contrary to the interests
of that group, then that government is removed or forced to
subordinate itself to that group.(81)
  It is not so important whether that interpretation is basically
accurate, i.e., whether Marx really claimed something like
that. What are essential are the methodological assumptions
which made Malewski adopt such an interpretation. Now the
assumptions obviously corresponded to the principles of
logical-positivist methodology in the earlier period, but are at
variance with the principles in the later period. Let us use
that imprecise classification for convenience, for all we mean is
this:

Malewski's interpretation followed the principles of logical-positivist methodology before the breakthrough which consisted in treating theory as a basic methodological structure, and which was reflected in, among other things, conception of theoretical constructs as terms whose meanings are determined by the assumptions of a given theoretical system. In the light of more recent (in this sense) ideas, Malewski's interpretation is inadmissible, because the operation is made on isolated parts of a certain theoretical system, without reference to the basic assumptions of that system. Malewski acted on the assumption, which was already then being questioned in writings on methodology, that empirical meaning is an attribute of separate statements, and this was why he set himself the task of separating, within the theory of historical materialism, those parts (understood as single statements or sets of statements) which are empirically meaningful from those which are not. This approach is, of course, set in a broader context which includes an idea which is very strongly marked in Malewski's works, of a cumulative development of science.

In the history of sociological thought we can single out the history of ideologies and the history of discoveries. If we disregard ideology and strive to bring out, with maximum benevolence, from the writings of the various thinkers, those general intuitions about close relationship which seem to agree with the now known facts, we obtain a rich collection of general hypotheses. These hypotheses very often cumulate to form the content of the achievements of science to date. New research can modify some of these hypotheses and add others. The work of the various researchers and of the various generations of researchers can be stratified in this way.(82)

It is disputable how far an objection which can easily be raised against other sociologists who adhere to logical-positivist methodology, namely the objection that they lag behind the evolution of the philosophy of science and import obsolete ideas, is applicable to Malewski. But it could hardly be denied that because of the very strong influence logical positivism had upon him, Malewski resorted to interpretative operations which it would seem every social scientist must intuitively reject as inadmissible.

We saw the impact of logical positivism upon sociology for theory 'construction' and 'operationalization' of concepts, but this impact could equally well be traced in any sphere of the sociologists' research activity, and we would see the same characteristic ways of adapting ideas that had developed in a different intellectual atmosphere. The essence of the problem is not that definite conceptions, methodological ideas and technical suggestions are taken over from the philosophy of

science; the point is that logical-positivist influence resulted in
an integrated methodological reorientation in sociology, manifested in theoretical studies and in the field work, in handbooks
of methodology and the choice of problems to be investigated,
in the techniques of data collection and in the criteria of academic achievements, in curricula and in the organization of
research work, in a word, in the totality of the sociologist's
research work.

The belief that all this belongs to the dead past would quite
simply be a fundamental error. As has been mentioned, the last
twenty years saw a marked increase in the philosophical competence of sociologists, especially of the leading group. The
new generation of handbooks of methodology and other texts,
in that respect represents an incomparably higher level than
the works of Zetterberg and Homans and handbooks like
'Methods in Social Research' by Goode and Hatt. The knowledge
which a present-day handbook of methodology conveys to
students of sociology is no longer a mixture of everyday ideas
about the natural sciences and clumsily formulated principles
of logical positivism from the period of the Vienna Circle. It
is usually a fairly correct though rather amateurish(83) synthesis of logical-positivist doctrine from, say, the early 1960s.
Rank-and-file sociologists still lag far behind that level of
methodological 'sophistication', but those texts set up new
standards of methodological knowledge. We must assume that
those standards will determine the development of sociology
over the next ten or twenty years.

6 Example I:
Paul Lazarsfeld: from concepts to indicators

Modern sociology has, roughly speaking, a double genealogy.
On the one side, it originates from the great theoretical systems
formulated in the nineteenth and the early twentieth century,
and on the other, from the tradition of empirical research,
usually linked to social and reform activity. Combining these
two traditions within one discipline proved possible and advan-
tageous, but it has not resulted in their full integration. The
various periods, schools and centers of sociology are usually
much more strongly connected with one of these two traditions
than with the other. A good example is offered, on the one hand,
by the Chicago School which, for all its significant theoretical
achievements, was primarily a school of researchers who took up
the urgent problems of urban communities,(1) and, on the other
hand, by the Frankfurt School which, for all its significant
achievements in empirical research, is not without reason known
primarily for its theoretical activity connected with the Marxist
system.(2)

Nothing would seem more natural than a full synthesis of these
two traditions: as is commonly known, theoretical reflection
needs to be supported by empirical research, and empirical
research should be linked to theoretical thinking. But it is not
as simple as that. Theoretical systems do not emerge in an
empirical vacuum, they are not products of pure speculation
(even though they are often presented as such) and, when it
comes to those systems to which modern sociology refers, they
have a strong foundation in empirical data, with the proviso
that the manner of collecting those data and their links with the
theoretical views usually do not meet the criteria of standard
methodology. On the other hand, empirical research is supported
in theory, but with the proviso that it is less a theory clad as a
system, than everyday social knowledge: the problems taken up
within the empirical tradition precede research activity, which is
to say that they are identified independently of research work -
one might say (if one admits such language) that they are
'imposed by life itself'. What comes in question then is not so
much integration of empirical research with theoretical reflection,
as integration of two trends in cognitive activity, each trend
being governed by its own system of rules of research procedure,
transmitted from generation to generation, developed and
improved.

Improvement here of course consists in adjusting research
procedures to the problems studied. And this is a very essential

issue. For whatever one might say, the traditional trend of empirical research was marked by a relative simplicity of theory. It included mainly survey-type studies, that is to say, descriptions of social facts, which in principle did not require any far-reaching theoretical background. Of course, survey-type research is not free from theoretical problems, not to speak of methodological ones, and sometimes requires immense ingenuity. Nevertheless, in comparison with the theoretical problems we face, for instance when investigating the class structure of a given society,(3) we are dealing here with relatively simple issues with problems we can solve on the basis of common sense and current knowledge, without engaging in problems which are theoretical par excellence.

Can the research techniques developed in connection with such research activity be successfully used in research that takes up intricate theoretical problems? The question is not easy to answer although we may assume that any answer that would settle the issue generally would be a simplification. In any case, the problem is there, and if so, then the use of those techniques as the techniques of sociological research would be based on questionable assumptions. I think that from that point of view we have to look at 'empirical sociology' in general, and at the methodological suggestions advanced by Paul Lazarsfeld in particular.

In the introduction to 'The Language of Social Research' Lazarsfeld writes that 'as a result of modern positivism, interest in clarifying the meaning of concepts and statements has become quite general,(4) and he sees his publications in that perspective. In fact, however, he means a special kind of clarification of meaning, namely the explication of concepts, in this case sociological ones, 'in the language of modern research procedures'. Such explication in practice means that 'we give up certain connotations [of the term involved] in order to make the remainder more precise and more easily amenable to verification and proof. From that point of view,

> It is instructive to examine the work of a classical writer, say one in the field of public opinion research, and to see *how his statements might be translated into the language of modern research procedures*. It will be found, on the one hand, that such writings contain a great richness of *ideas which could be profitably infused into current empirical work;* on the other it will be found that such a writer tolerates great ambiguity of expression. The task of such explication is not to criticize the work, but rather *to bridge a gap, in this case between an older humanistic tradition and a newer one which is more empirically oriented.*(5)

The more complex a given concept, Lazarsfeld says, the more we need explication understood in this way, that is, *translation of concepts into observable indicators.* Hence the importance

which Lazarsfeld attaches to the explication of concepts originat-
ing in 'an older humanistic tradition', or, as I would prefer to
call it, in theoretical systems.

How are concepts translated into observable indicators, or, in
other words, how are observable indicators selected to fit the
concepts? Now, 'this process by which concepts are translated
into empirical indices has four steps: an initial imagery of the
concept, the specification of dimensions, the selection of observable indicators, and the combination of indicators into indices.'(6)

The first of these operations can best be illustrated by an
example.

> Suppose we want to study industrial firms. We naturally want
> to measure the management of the firm. What do we mean by
> management and managers? Is every foreman a manager? Some-
> where the notion of management was started, within a man's
> writing or a man's experience. Someone noticed that, under
> the same conditions, sometimes a factory is well run and some-
> times it is not well run. Something was being done to make
> men and materials more productive. That 'something' was
> called management, and ever since students of industrial
> organization have tried to make this notion more concrete and
> precise.(7)

The next operation consists of breaking down a given concept
into 'aspects', 'components', and 'dimensions'. It then turns out,
Lazarsfeld says, that a concept is more a combination of
phenomena than a simple, directly observable phenomenon. Such
a breaking down of a concept into its components should not go
too far, but on the other hand, 'as a general principle, *every
concept we use in the social sciences is so complex that breaking
it down into dimensions is absolutely essential in order to translate it into any kind of operation or measurement.*'(8)

What Lazarsfeld calls selection of indicators, in fact, covers
operations of two kinds: 'thinking up' the indicators, which
yields a list of many possible indicators of a given 'compound'
property, and the choice from among them of those which are to be
used in future research. In practice, these two operations are
carried out separately for every 'dimension' of a given concept.
In other words, a 'battery' of indicators is assigned to every
'dimension'. The selection of indicators requires, among other
things, a decision to settle 'which indicators are considered
"part" of the concept, and which are considered independent or
external to it.'(9)

Such far-reaching breaking down of a concept into its components makes it inoperative. Hence all the elements have to be
reintegrated, and this is the operation called index construction.

> After the efficiency of a team or intelligence of a boy has been
> divided into six dimensions, and ten indicators have been
> selected for each dimension, we have to put them all together,

because we cannot operate with all those dimensions and indicators separately.(10)

At this point a serious difficulty emerges. The various indicators either cannot be measured at all or are measured with different incomparable scales, so that there is usually no formula that would establish the weight of the various indicators Then how can the indicators by 'put together?' This is an age-old and perfectly well-known problem, which in this case makes Lazarsfeld pose a fundamental question: 'Can you really develop a theory to put a variety of indicators together?' He thinks we can, in principle, answer that question in the affirmative.

The subject is a large one, and it is impossible to go into detail here. The aim always is to study how these indicators are interrelated with each other, and to derive from these interrelations some general mathematical ideas of what one might call the power of one indicator, as compared with another, to contribute to the specific measurement one wants to make.(11)

It is difficult to say what mathematical operations he had in mind. Perhaps he meant the latent structure analysis, worked out by him in connection with Stouffer's studies of the American army. If so, objections arise. Latent structure analysis has not been fully worked out.(12) This is perhaps why the application of that idea has not advanced beyond experiments, and it would be difficult to say that that idea solves the problem. Further, as Stefan Nowak writes, latent structure analysis is 'a method of clarifying interrelations among answers by reducing them to a latent common factor.'(13) Thus it leads to answers to questions about the interpretation of given results of empirical studies (not necessarily questionnaire studies), and not to a formula that would enable one to put together a given set of indicators so as to obtain a pertinent index of the concept being analyzed. Latent structure analysis is thus rather a method for constructing substantially grounded concepts than for empirically reconstructing concepts shaped in a different way, unless we mean reconstruction in a very loose sense, namely an operation in which the existing conceptual apparatus is a nonobligatory point of departure and performs a solely heuristic function.

Contrary to Lazarsfeld's optimistic assurances, I think that there is no general answer to the question how indicators are to be put together to form an index, and the problem is solved from case to case on the basis of common sense, which yields better or worse results primarily depending on the concept with which we are dealing. In each case, more or less arbitrary decisions of necessity play an essential role, and it can easily be seen that the more complex the concept with which a given author is dealing, the more arbitrary he is in manipulating the indicators. There are, it seems, certain unclear but more or

less impassable limits to reasonable synthesizing of indicators.
Once that limit is passed we enter the area of practically complete arbitrariness.

Whoever has had to deal with empirical sociological studies knows
how many interpretations are required even in the case of the
simplest concepts. 'Education', 'income', 'occupation', 'household' and similar terms usually create no problems in everyday
conversation because the context of that conversation suffices
for interpreting their meaning. But research forces one to
import much greater precision to them (e.g., is 'sociologist' an
'occupation?') and to standardize the set of the concepts used
(if the data are to be comparable). This is usually followed by
regulating the meaning of a given term on a semantic or factual
basis. For instance, it is not an indifferent issue (from the
purely cognitive point of view, not to speak of other viewpoints)
whether in present-day Poland the advantages resulting from
obtaining a special document that entitles one to purchase a car
at the official price are to be included in a person's income.
 Such interpretative operations often follow the course described
by Lazarsfeld. Note, however, that this is so only if the
researcher enters a theoretically unexplored area. In such a case
the whole undertaking begins with a term and reflection on its
meaning (which is, of course, not to say that the further procedure must be in agreement with that advocated by Lazarsfeld).
If, however, a person engages in the study of the organization
of an enterprise or labour productivity (examples quoted by
Lazarsfeld), then he enters an area of comparatively intense
research activity, which yields better or worse patterns of
research procedure and, most important of all, some theoretical
achievements. Entering that area forces him to adopt an attitude
toward those achievements, that is, to take up more or less
advanced theoretical issues. Can he then successfully follow
Lazarsfeld's recommendations? Theoretically he can, but in
practice this is unlikely. It is not very likely that fairly
expanded theoretical problems should be reducible to several
aspects and several dozen indicators.
 The case of the Geneva method of measuring the standard of
living is instructive from this point of view.(14) The procedure
used by the authors (a team of researchers headed by Jan
Drewnowski) strictly corresponds to that advocated by
Lazarsfeld, which in this case was due not so much to methodological borrowing as to the impact of the same pattern of
empirical studies. The concept of the standard of living was in
the course of analysis broken down into seven dimensions corresponding to the groups of basic needs. Three or four representative measures were then assigned to each group of needs
(each dimension); use of those measures yielded numerical
results which formed the basis for computing a synthetic index
of the standard of living. Unfortunately, the method does not
withstand substantive criticism: it refers to a sequence of

arbitrary decisions; the list of the basic needs, the choice of
representative measures, the establishment of threshold values,
the assigning of weights - all these operations are performed in
a theoretical vacuum.(15) And this is not to be wondered at. The
problems of the standard of living, as we see them today, are
very complex, and are closely interwoven with purely theoretical
issues of social progress. Taking up such problems on the basis
of methodological patterns of simple social statistics necessarily
results in merely apparent successes.

Lazarsfeld constructed his pattern of research procedure with
a certain type of social survey in mind. On the basis of that
pattern he laid the foundations for modern surveys. As it turned
out, the pattern is very fertile. But the point is that for
Lazarsfeld it was a universal pattern, also applicable in situations
where we deal with complex theoretical problems. Lazarsfeld set
himself the task of combining the tradition of empirical studies
with that of great theoretical systems. Bringing different traditions together often kills one of them. That was true in this
case, too. Lazarsfeld suggested a synthesis based on the conditions set by the tradition of empirical studies. Theoretical
conceptions are assigned a heuristic function: they are a good
source of inspiration in research, but research proper begins
later, beyond those conceptions, with the process of 'translating'
them into 'the language of modern research procedures'. It is
no coincidence that on one occasion Lazarsfeld wrote about
bridging the gap between 'an older humanistic tradition and a
newer one which is more empirically oriented', and another time,
about making 'a science out of the more general and much older
attempts to understand human society'. It is also known from
other statements made by Lazarsfeld that in his eyes the science
of society was in its very beginnings. Its genealogy reaches as
far back as the tradition of empirical research that corresponds
to the above-described principles of research procedure. The
rest, i.e., primarily the theoretical systems, is the pre-
scientific (extra-scientific, non-scientific) penumbra of nascent
scientific sociology. Lazarsfeld knew the theoretical achievements
of sociology well and held them in high esteem, but he saw other
values in them than those he associated with the concept of
science. This obviously had definite consequences when it came
to constructing the pattern of scientific procedure.

That pattern was, of course, not given an openly normative
form. Lazarsfeld certainly was one of the most enlightened
methodologists and he saw the limits of admissible interference
with research practice on the part of methodology. 'The methodologist is a scholar who is above all analytical in his approach
to his subject matter. He tells other scholars what they have
done, or might do, rather than what they should do.'(16) Unfortunately, what Lazarsfeld actually described, qua methodologist,
was not the research practice of sociologists, but research
procedure in sociology, or in other words, the procedure used
in scientific sociology; furthermore, he did not write how con-

cepts happened to be translated into the language of research operations, but how a concept is translated into the language of research operations. The normative sense of such methodological statements was perhaps realized neither by the author nor by his readers, but in the case of Lazarsfeld's publications, especially 'The Language of Social Research', their standard-setting effect is beyond dispute.

Be that as it may, the above-described procedure of transition from concepts to research operations has been accepted in the main trend of sociology as the general schema of research procedure in sociology - in sociology in general, and not just in surveys or in research concerned with specific problems. In other words, a certain type of social survey, improved and expanded, has been assigned the status of sociological research plain and simple.

One can easily notice the resemblance of the procedure of the selection of indicators in Lazarsfeld to the operational definition in Zetterberg. Lazarsfeld in fact in slightly different language expressed the same idea which Zetterberg stated on the basis of the concept of operational definition. The procedure of passing from a concept to indicators is the procedure of translating a concept into the language of 'research operations', defined as such by the tradition of empirical research.

In Lazarsfeld's opinion, operationalization understood in this way was indispensable. 'In order to incorporate the concept into a research design, observable indicators of it must be selected.' It is true that in this case Lazarsfeld did not go beyond the tautological (in his language) statement that if one wants to incorporate a given concept into a research process at the level of 'observable phenomena', one has to translate that concept into the language of 'observable phenomena', but for the people who use them such latent tautologies have the force of absolutely binding (because evidently correct) requirements. On the other hand, it is not clear whether, according to Lazarsfeld, that requirement should be given any epistemological interpretation. In particular, it is not clear whether for Lazarsfeld the selection of indicators was a case of assigning an empirical meaning to a concept. Lazarsfeld having been a very pragmatically oriented methodologist, it seems that such an interpretation would be unauthorized. What he wrote at most entitles us to suppose that he treated the selection of indicators as an operation of assigning a meaning to terms which was interesting in his research perspective (which he identified with the research perspective of sociology). It is claimed, therefore, that Lazarsfeld imposed his version of the requirement of operationalization not because of epistemological considerations, but because it followed from his idea of the nature of the facts with which sociology is concerned.

The requirement of the selection of indicators for concepts, however, was given epistemological sanction in the works of those of Lazarsfeld's disciples who were more sensitive to philosophical issues. It can be found in its most mature form in Stefan Nowak:

> If we do not know the empirical meaning of the terms we use we are unable to subject the hypotheses we have formulated to an empirical verification, because we do not know which phenomena should be observed to validate our hypothesis, and which should be observed to refute it. Nor are we in a position to classify phenomena in terms of our conceptual apparatus, because the operation of classification requires a previous observation of the criteria on which that classification is based. Further, we are not in a position to predict the occurrence of the various phenomena, because in order to predict the occurrence of a phenomenon B it is necessary to first identify a phenomenon A which is followed by B either in every case or with a definite probability. *That in turn requires that the terms in our theory or the concepts we use in analyzing phenomena, should have empirical meaning, or, as others put it, that they should be defined operationally, reductively, or, in still other wording, that they should have appropriate empirical indicators.*(17)

It follows from the footnote that the operational defining of concepts, reductive defining of concepts and selection of indicators are treated by Nowak as 'various methods of assigning an empirical meaning.' But on the whole, people influenced by Lazarsfeld do not explain those distinctions and more or less consciously identify the idea of the selection of indicators with the idea of an empirical interpretation of terms.

7 Example II: the methodology of comparative studies

Ten years ago James Coleman summed up his comments on the state of the methodology of comparative studies by stating that 'There has not appeared a literature on the specific methodological problems of cross-national surveys; it is clear that one is coming to be needed.'(1) We have since then seen a number of publications that took up methodological issues of comparative studies, and surveys in particular.(2) It is difficult to say whether the achievements of these ten years would make Coleman change his above-mentioned opinion, but as I see it, such a change would not be justified. The methodological literature concerned with comparative studies is fairly comprehensive when it comes to the number of publications and fairly advanced as compared to other branches of the methodology of social research, but it does not in any essential way solve the methodological problem of comparative studies, and sometimes even does not pose that problem, thus ignoring the issue of its own reason for being.

'In mathematics,' says Sjoerd Groenman, 'and also in physics and chemistry, data are not nation-specific. They can be studied in any surroundings with the proper tools. In the social sciences, however, phenomena are context-bound.'(3)

Groenman obviously means the well-known fact that if, when observing a social phenomenon, we move from one country to another, in most cases we find that the phenomenon must be studied in the context of a given country if it is to be understood properly. Political behaviour must be studied in the context of the political system in a given country, because it is that context which imparts it its proper meaning; a comparison of voting behaviour in Poland and in the United States, not based on a contextual interpretation of the data, would be a completely pointless undertaking.

These things are known not only to sociologists who try to make cross-national comparisons, and on the whole not only in connection with cross-national comparisons. Historians have long since pointed out the significance of the historical context for a proper understanding of historical facts, by which they meant, among other things, the context of a given historical epoch. For a social anthropologist the context of culture understood anthropologically is an essential thing. Any Polish sociologist who before 1939 studied the religious behaviour of the Polish rural population knew how important the context of the peasant culture was for the comprehension of that behavior.

It seems therefore, generally speaking, that social facts happen to be linked in a peculiar manner to various social systems (culture, historical epoch, nation, local community, vocational milieu, etc.): they are linked to them in such a way that the systemic context determines the character, or to put it more strongly, the nature of those facts. This state of things is reflected in the conceptual apparatus in the social sciences, a part of which at least has the character of historical concepts, i.e., ones specifically connected with a definite system (historical epoch, historical moment, culture, etc.). The meaning of that connection is that the application of a given historical concept outside the boundaries of a given system is not so much an ordinary factual error as a violation of language rules (which may, and usually is, due to a lack of sufficient knowledge of facts, but that is another issue). The concept of serf is logically connected with that of feudalism, and anyone who uses it with reference to present-day Latin America, for instance, either uses it metaphorically or fails to understand its meaning.

The historical character of the concepts the sociologist deals with is not always equally clearly visible. Consider the concept of the intelligentsia. It was shaped in Eastern Europe and has been closely connected with the history of that region. Hence its use is justified only within certain historical limits determined by the temporal and territorial scope of the occurrence of that complex of features which marked Eastern European societies in the second half of the nineteenth century and in the first two or three decades of the twentieth century. The use of that term with reference to any social category or group outside those - very vague, it is true - limits is often just a confusion of concepts. It is a debatable point whether we can speak about the German intelligentsia, whatever period we have in mind, or about the intelligentsia in postwar Poland, but there is no doubt that if someone talks about the American intelligentsia he evidently identifies the intelligentsia with the intellectuals or with some other category of educated people, and thus changes the meaning of the term under consideration. It is a fact that the intelligentsia can be distinguished in its social milieu by a combination of two 'universal' characteristics: education and its availability on the labor market, but, on the other hand, people marked by these two features were turned into members of the intelligentsia by the specific socio-economic conditions that prevailed in nineteenth century Eastern Europe. They were turned into people who occupied a specific place in the social structure, were marked by a specific style of life, played the role of spiritual leaders of the nation, etc. To put it briefly, the intelligentsia was a product of a certain, historically localized, and hence unique, complex of social conditions, and it can be defined only against the background of those conditions. It is also an element of a certain social whole whose comprehension requires comprehension of that whole.

The problem of comparative studies can accordingly be formulated thus: the meaning of certain social phenomena being to a large extent determined by some historical context or other, what kind of research procedure enables one to avoid erroneous interpretation, or at least reduces the risk of error, in comparisons which cross important contextual boundaries? The very fact that this question is posed assumes that the contextual involvement of social facts is a methodologically important problem, i.e. that it creates specific difficulties in the process of research. On the other hand, it is assumed that those difficulties can be surmounted. The first assumption is fairly obvious. If one wants to compare social phenomena, one has to find a common denominator for them, and that is certainly a problem in situations in which the meanings of those phenomena are determined by essentially different contexts. The second assumption is a postulate useful in practice: as long as there is no sufficient proof that a given problem is unsolvable, one has to assume that it does have a solution. The rejection of either assumption puts an end to the idea of comparative studies as something either trivial or misconceived.

What then do comparative studies consists of? Of course not of a mere comparison of facts, nor even of a comparison of facts from different systems (whatever that could mean). We can speak about comparative studies only when the comparison brings up those specific methodological problems which result from the contextual involvement of the facts compared. From that point of view, the difference between the various cases of comparative study is one of degree, but for practical reasons it is worth singling out those situations in which contextual involvement of phenomena is a serious issue. Usually it is quite correctly assumed, that such situations occur in the case of cross-national and cross-cultural studies. This is not to say, however, that all cross-national studies are automatically to be treated as comparative studies. A comparison of voting behavior in Poland and Czechoslovakia would be practically free from the problems we are talking about. On the other hand, many comparisons made within one and the same country require 'the comparative approach'. Generally speaking, whether a given research project is in the sphere of comparative studies depends on the kind of problems we are dealing with and how those problems are solved. Any general criterion, based on some classification of social systems, leads to artificial divisions, as shown, for instance, by the following definition advanced by Marsh:

> Comparative sociology may be defined as that field which is concerned with the systematic and explicit comparison of social phenomena in two or more societies. Cross-societal (cross-cultural, cross-national) comparison is the essential ingredient; intra-societal comparisons may or may not be made concurrently with cross-societal comparisons. Studies which are clearly excluded from comparative sociology as defined are those which make intra-societal comparisons - e.g., between

middle and working class voting patterns in one society -
without also making cross-societal comparisons.(4)

Why should political frontiers be more important from the
methodological or factual point of view than social frontiers? The
reason why comparative studies exist as a specific category of
studies and comparative sociology as a subdiscipline is that some
comparisons are particularly difficult to make and entail specific
problems. If such problems emerge primarily in connection with
cross-national and cross-cultural comparisons, then this is a
useful practical indication, and not a basis for singling out cross-
national and cross-cultural comparisons as a separate category of
studies.

If the methodological problems of comparative studies are connected with the fact that the contextual involvement of social
phenomena makes them difficult to compare, then an analysis of
that fact should be the starting point for the methodology of
comparative studies. Let us then see how far this is really so.
Now some methodologists of comparative studies fail to notice
this problem at all. For instance, Marion J. Levy, Jr., thinks
'it is something of an amusing scandal that among the contributions to a volume on theoretical sociology there should be a
special paper on the subject of comparative analysis.' It is a
'scandal' because 'all scientific analysis is a subset of the general
set entitled comparative analysis!'(5) A similar opinion is held by
Morris Zelditch, Jr. In his eyes the differentiation into comparative and non-comparative sociology is 'arbitrary and without
real methodological significance' because 'all explanatory,
generalizing research involves comparison, and almost every
known research design in sociology is therefore comparative,
whether experimental, survey or macro-sociological.'(6)
In this trivial sense not only 'all explanatory, generalizing
research' but all existing and possible research is comparative.
No analysis is possible without comparison, and in this sense all
thinking is comparative, too. Basing himself on these assumptions, Zelditch comes to the conclusion that the only reason for
which comparative studies and comparative sociology are singled
out is an 'unreasonable convention', which he nevertheless does
respect sufficiently to formulate 'the logical foundation of comparative analysis', which consists of the following four rules:

(1) (Comparability.) Two or more instances of a phenomenon
may be compared if and only if there exists some variable,
say V, common to each instance.
(2) (Mill's 1st Canon.) No second variable, say U, is the
cause or effect of V, if it is not found when V is found.
(3) (Mill's 2nd Canon.) No second variable U is the cause or
effect of V if it is found when V is not.
(4) (Rule of One Variable.) No second variable U is definitely
the cause or effect of V if there exists a third variable, say W,

that is present or absent in the same circumstances as U.(7)

In my opinion, these rules could easily be questioned both from the logical (rules 1 to 4) and the historical (rules 2 and 3) point of view. But this is not the core of the matter. The point is that 'the logical foundation of comparative analysis', as suggested by Zelditch, has all the characteristics of a pidgin logic, with emphasis on logic rather than on pidgin: the essential issue is not so much the fact that the logic in this case is very elementary and far from correct, but rather that there is nothing except logic here. None of these four rules has anything to do with the real problems of comparative studies and they do not function as the logical foundation of comparative studies any better than any other set of theorems drawn from a handbook of elementary logic. And yet Zelditch treats them almost like the axioms of a serious theory of comparative studies, which enables him to score easy victories in disputes with his adversaries:

> Many ethnologists believe that our understanding of a society is violated if 'native meanings' are not presented in comparison. This clearly conflicts with the implications of rule 1. It may therefore be regarded as 'unsound'; and nothing more clearly reveals the central importance of rules 1-4 than the role they are able to play in ruling out unsound rules.(8)

Zelditch goes to extremes in his opinion, but his views are interesting in showing clearly the process of the emergence of what in a previous chapter was termed the external methodology of social research. Zelditch completely disregards the real methodological problems of comparative studies by dismissing both theoretical reflections on the historical variability of social phenomena, their cultural differentiation, etc., and the experience of a now quite large number of people who specialize in comparative studies. His methodological model has practically no connection at all with research practice, which obviously does not change the fact that it can affect that practice. On the other hand, his case is typical in that most publications dedicated to the methodology of comparative studies start from certain assumptions adopted in advance, and even if they take up real problems of comparative studies they do it so as not to infringe those assumptions. In more concrete terms, the methodology of comparative studies, as presented nowadays, is an attempt to apply the standard 'methodology of social research' to problems which have emerged in connection with cross-national comparisons. Hence even the best publications in that field adjust the problems to the model adopted from the outside rather than adapting the model to genuine research problems.

Most methodological writings concerned with comparative studies pertain to technical issues connected with cross-national comparisons, such as sampling, organization of field work, translation of questionnaire items, training of interviewers, etc.

The standard of these publications has risen considerably durinɡ
the last ten years, mainly thanks to the fact that under the impa
of a rapid spread of comparative studies, discussions of technica
issues have turned into an exchange of experience in research.
But nevertheless, these publications avoid posing basic methodo
logical problems. Their authors implicitly assume the applicabilit
of the standard 'methodology of social research' to comparative
studies and try to solve the problems which arise when that
methodology is actually being used. Thus, without considering
the applicability of surveys in comparative studies, they have
been posing the same question for years, unfortunately, always
with the same results: how to minimize, or at least bring under
control, those factors which contribute most to the unreliability
of the results of surveys in comparative studies (incomparability
of samples, differences in the definitions of the fundamental
'variables', incompetence of many local research teams, etc.).

The second category of publications on the methodology of com
parative research covers the few items in which the basic problems of comparative studies, connected with the contextual
involvement of social phenomena, are being posed, and endeavor
are made to solve them. One of the best (if not simply the best)
publications of this kind is 'The Logic of Comparative Social
Inquiry' by Przeworski and Teune.(9) Its authors evidently com
bine genuine experience in research with a high standard of
logic and knowledge of general scientific methodology. Let us
see how they formulate and solve the problem of comparative
studies.

Unlike Zelditch, Przeworski and Teune realize the significance
of the contextual involvement of social phenomena. They are
critical of those comparative studies which 'have accepted a
methodology validated by social science practice in single culture
particularly in the United States' and

> have tended to view the problems of comparative research in
> terms of research difficulties, such as translating question-
> naires, training interviewers, assuring accuracy of data, and
> have tended to ignore the insight of the area specialist: a
> nation, culture, or region must be considered as a 'whole'.(10

By postulating that the various social systems be treated as
wholes, Przeworski and Teune suggest a research strategy that
would take into account the methodological consequences of such
an approach.

It turns out, however, that the new research strategy advocated by Przeworski and Teune is not, and cannot be, essentiall
different from the strategy they criticize because the opinions
as to what is and what is not science, on which they base themselves, are such that any considerable deviation from the
standard methodology of sociological research is out of the
question.

They start from the assumption that 'the goal of social science is to explain social phenomena,'(11) 'explanation' being defined in accordance with Hempel's model. That model, as we know, assumes that to explain a fact is to indicate the appropriate general law (or set of general laws) and the initial conditions. A general law is understood as a universal statement, i.e., a statement which says that 'universally, if a certain set of conditions, C, is realized, then another specified set of conditions, E, is realized as well.'(12) Thus, in order to explain a fact, we have to refer to a general law in the form of a universal statement.

The attainment of the goal of the social sciences, as defined by Przeworski and Teune, implies looking for general laws in the logical-positivist sense of the term. Such a definition of the goal of the social sciences has definite consequences; in particular, it forces one to direct comparative studies accordingly. This issue will be discussed a little later. First, we will ask, why the authors thus defined the goal of the social sciences and whether that goal is attainable.

This question is based on premises of two kinds. First, no one has yet demonstrated that general laws as understood in logical positivism, i.e., in the form of universal statements, are theoretically more valuable than historical laws or what might be termed laws of culture, that is, general statements concerned with a given historical period (e.g., historically interpreted capitalism) or a system of culture (e.g., Western civilization). A statement which refers to capitalism (as historically localized socio-economic formation) need not be either a singular statement or a statement reducible to singular statements or a historical generalization (understood as a statement with spatio-temporal co-ordinates). There is something like a theory of capitalism, and if a person associates it with a set of historical generalizations (i.e., statements with spatio-temporal co-ordinates), he probably does so because capitalism can be localized in time and space. But if we examine theories of capitalism (like those of Karl Marx and Max Weber), we see that the spatio-temporal localization of capitalism is not a definitional thesis in them. If this suffices to class that thesis as an empirical one, then it should so be classed. This is, however, a debatable point. Looking at the theory of capitalism for empirical statements, on the one hand, and for statements which explain the concept of capitalism, on the other hand, is a risky undertaking. 'Capitalism' is the name of a certain whole which is defined not so much by a definition, a set of postulates, rules of correspondence or anything like that, but rather by a given theory of capitalism as a whole. It is not legitimate to single out from among its statements those which explain the meaning of the term capitalism or the range of the phenomena which belong to that whole and those which state what the properties of capitalism understood in this way are. A statement which localizes capitalism within certain (quite vague) temporal and territorial limits, co-defines capitalism in a very loose sense, but it is also an empirically verifiable statement on the limits within which one can deal with capitalism. The statements which belong to the theory of capitalism are thus neither universal statements (because they apply to a historically localized whole which cannot be described in universal terms) nor historical generalizations (because that whole is historically localized by the totality of our historical knowledge, and not by spatio-temporal co-ordinates; the temporal and territorial limits of that localization are a not very important element of that localization and every statement about them is subject to empirical verification).

The premises of the second kind pertain to whether the program of looking for universal laws in the social sciences is a realistic one. It is instructive that the conscious search for such laws, for a hundred years or more, has not yielded results that would encourage us to continue the search. Examples of universal statements in the social sciences, quoted several times,(13) fall under one of the following three categories:

(i)   trivial statements;
(ii)   statements which are evidently false in the light of
       historical and anthropological data;
(iii) statements which Ossowski included in social zoology,
       i.e., statements referring to properties which man shares
       with at least part of the animal world.(14)

The last category of statements is important and interesting, but it cannot replace specifically humanistic knowledge.

Why then do Przeworski and Teune define the goal of the social sciences so as they do? Their reply is:

The pivotal assumption of this analysis is that social science research, including comparative inquiry, should and can lead to general statements about social phenomena. This assumption implies that human or social behavior can be explained in terms of general laws established by observation. *Introduced here as an expression of preference, this assumption will not be logically justified.*(15)

This is a very interesting statement. As people engaged in theoretical reflection and having their own experience in research, among other things in comparative studies of authority in local communities,(16) Przeworski and Teune certainly realize that their assumption is, let us say, debatable. This is also clear from their above-quoted statement. And yet they do adopt that assumption. I think, though it would be hard to check, that they adopt it without justification because they know the social sciences and social reality too well to repeat the superficial arguments in favor of a program for constructing a universal theory in the social sciences, advocated by philosophers of science. But, on the other hand, they are unable to abandon stereotyped and deeply rooted beliefs concerning the goals of science, theory, explanation, etc. Such being the case, they choose to present their program as their own preference. But behind this preference is less an act of simple faith than a well-developed system of opinions known from somewhere else.

When they adopt that system of opinions as their point of departure, Przeworski and Teune are forced to look for a strategy of comparative studies which would bring those opinions into agreement with knowledge of the contextual involvement of social phenomena. If it is assumed that construction of universal theories in the social sciences is possible, one has to look for a way to define phenomena which have a definite historical or cultural setting, in universal terms. Since the meaning of a given phenomenon is determined by its systemic context, it must be assumed that that context can somehow be expressed in general terms. Moreover, the assumption that the goal of the social sciences consists in construction of a universal theory determines, in the light of our knowledge of the historical and cultural conditioning of social facts, the place and function of comparative studies in the social sciences. Comparative studies are accordingly treated as an operation which makes it legitimate, in a sociological description, to cross the boundaries of various cultures and epochs, so as to eventually formulate universal laws. As Marsh put it, 'The fundamental rationale of comparative sociology is, then, the need to universalize sociological theory.'(17) In such approach, comparative studies become part of the programme of 'theory construction and verification' i.e. construction and verification of a system of universal statements.

And this is how Przeworski and Teune see comparative studies - as an important step in the process of construction of a universal theory, and the importance of that step is due to the fact that 'particular social systems do influence the nature of observed relationship and do yield a gain in prediction.'(18) This is why research procedure that strives to construct a universal theory consists mainly of establishing what in a given system affects the nature of observed relationships. Hence, 'the goal of comparative research is to *substitute names of variables for the name of social systems,* such as Ghana, the United States, Africa, or Asia.'(19) This is to say that 'historical statements are implicitly theoretical. They subsume under the proper names of the social systems a broad range of factors that might be used in theoretical explanation.'(20) Thus, in a sense, historical statements play the role of useful hints as to where important explanatory factors are to be looked for.(21)

In the programme formulated by the authors as a result of these considerations, it is assumed that the names of nations and other social systems should be treated as 'residua of variables that influence the phenomenon being explained but have not yet been considered.' The programme suggests that such names be replaced by 'variables'.(22)

Is such a programme feasible? If one admits as Przeworski and Teune have, that the comparability of social phenomena is limited by their systemic connections, is it realistic to assume that the names of systems can be replaced by 'variables'?

Their answer is strongly in the affirmative. They say that

'comparability' depends upon the level of generality of the language that is applied to express observations. The respons to the classical objection to comparing 'apples and oranges' is simple: they are 'fruits'. And the answer to the question whether there are interest groups in the Soviet Union depends upon the level of generality of the concept 'interest group'. (2)

Of course, we can compare whatever we like with whatever else we like, provided we do so at a sufficient level of generality But the higher the level of generality the smaller the amount of information conveyed. Let us revert to the case of the intelligen sia. The answer to the question whether the intelligentsia is, or was, to be found in the United States, or in medieval France, or among Azande people, depends in a sense upon the level of generality of the concept of the intelligentsia. One can, of course, define 'the intelligentsia' so that the term loses all of its Eastern European connotation. But what is the advantage of that? In fact, it is just that Eastern European connotation which is decisive for the cognitive value of that concept. If we define the intelligentsia in terms of education (but is education a concept that is independent of its historical and cultural context?), then we obtain a different term for educated people. But is that what we have in mind?

Przeworski and Teune would probably reply as follows: If the historical connotation of 'the intelligentsia' is really so valuable, then we can try to preserve it. The point is not to abandon the systemic meanings of terms, but to universalize them (i.e., to change them into 'variables'). Hence, if we conclude that the intelligentsia is constituted by something more than just education and type of work, then we have to look for other 'variables' that combine to form that concept (in the process treating nineteenth-century Eastern Europe as 'the residuum' of the appropriate variables). We can then take the position of the intelligentsia in the social structure, the social functions and the political role of that group, etc., into account.

Perhaps in this particular case we could go quite far in that direction and in the end obtain an interesting set of 'variables.' Perhaps we would succeed, and perhaps not, or, to put it more precisely, perhaps the set of characteristics we would obtain would be substantially useful, or perhaps it would be totally artificial. Whatever the result, however, this set of characteristics would not fully correspond to the concept of the intelligen sia, unless it were interpreted against the background of the social, economic and political conditions which prevailed in nineteenth-century Eastern Europe, but that would amount to abandoning the 'universalization' of the concept of the intelligentsia. In other words, if we went in that direction, we would have to replace the concept of the intelligentsia by some poorer con-

ceptual construct. How much poorer would it be? This is, of
course, an open question, but we can certainly say that our
possibilities of understanding Eastern Europe would be
seriously limited. 'Universalization' of concepts can take place
only at the cost of their trivialization.

Everything seems to indicate that Przeworski and Teune would
answer: if so, then it cannot be helped.

> we . . . postulate that the generality and parsimony of theories
> should be given primacy over their accuracy. In other words,
> social science theories, rather than explaining phenomena as
> accurately as possible in terms relative to specific historical
> circumstances, should attempt to explain phenomena wherever
> and whenever they occur.(24)

This is, as we see, the next manifestation of preference, also
known from other sources. At this point, the possibility of
continuing the discussion ends. If the generality of a theory is
its primary value, then any argumentation intended to show that
other values are being lost in the quest for generality, misses
the point.

The above example is of particular interest to us, as it shows
how effectively a methodological doctrine can neutralize theoretical
knowledge and research experience. Przeworski and Teune comprehend well what the methodological problems of comparative
studies consist of. They have their own practical experience and
a good theoretical orientation in that respect, but their vision
of science makes them look for purely formal solutions to these
problems. As a result, these problems are not so much solved as
trivialized by them. For all practical purposes, they have given
a wide berth to what is really important and difficult to solve.
This, unfortunately, applies to the methodology of comparative
studies as a whole. Doctrinal limitations force people either to
ignore what we have here called the problem of comparative
studies, or to look for apparent solutions that follow from the
internal problems of a methodological doctrine adopted from the
outside, rather than from an analysis of the real obstacles encountered in the process of research. As a result, this
methodology leads research practice astray rather than helping
it. Such practice is pertinently characterized by Charles Taylor:

> the result of ignoring the difference in intersubjective
> meanings can be disastrous to a science of comparative politics,
> viz., that we interpret all other societies in the categories of
> our own. Ironically, this is what seems to have happened to
> American political science. Having strongly criticized the old
> institution-focussed comparative politics for its ethnocentricity
> (or Western bias), it proposes to understand the politics of
> all society in terms of such functions, for instance, as 'interest
> articulation' and 'interest aggregation' whose definition is

strongly influenced by the bargaining culture of our civilizatio
but which is far from being guaranteed appropriateness elsewhere.[25]

Postscript

This may sound absurd, but I would risk the claim that a good starting point for the renewal of sociological thought would be to abandon the privilege of using the term science. That privilege has unfortunately cost sociology too much. It is true that in some countries sociology is called a science just on the basis of linguistic usage, but in the English-speaking countries, where the profile of present-day sociology was shaped, that entering 'the great family of science' took place at the expense of sociology's losing its identity and, accordingly, its own cognitive abilities. Sociology would gain in value and - in the long run - would consolidate its prestige by adopting the provocative principle of Jacob Burckhardt: 'Wir sind "unwissenschaftlich" und haben gar keine Methode, wenigstens nicht die der anderen.'

This would not, of course, change the fact that sociology belongs to science as a certain trend in the history of culture, but it would help it to make itself independent of the pressure of current stereotypes that narrow down the concept of science to secondary correlates of research activity in natural science. Moreover, once the deck was cleared for action, the academic activity of a sociologist would have to be assessed on its merits, and not in the light of its agreement with the alleged picture of natural science, which happens and will continue to happen even independently of the logical-positivist influence. For too many authors brandishing the term science, scientific, etc., has become a defence mechanism that effectively protects them against factual criticism. By analogy, too many excellent sociological publications are ranked poor, or are even driven beyond the limits of academic sociology, only because the authors have disregarded the overwelming force of the requisites of 'being scientific'. In sociology 'being scientific' functions as an independent criterion of academic achievement. In practice that criterion is stronger than that of assessing a publication on its merits. If the criterion of 'being scientific' were eliminated, we might have to change the rankings of publications, research projects, authorities, etc., completely.

For the philosophers of science it is probably essential that the picture of sociology, as drawn here, and especially the picture of sociology's connections with the philosophy of science, more or less fits other social sciences as well. It is characteristic that it least well fits economics, which is little complex-ridden, and those disciplines, such as history, in which less importance

is attached to the status of being scientific than in sociology.
The comparatively strongest impact of the philosophy of science
can be seen in those disciplines which are related to sociology,
namely psychology, social psychology and political science.
Recently we can also watch a considerable effort on the part of
social anthropologists to bring their discipline close to sociology
in that respect.(1)

It is essential, however, that in every humanistic discipline
'scientific' methodology finds some echo and active advocates.(2)
Not to confine ourselves to mere claims, we shall refer to a
characteristic example. J. Rogers Hollingsworth thinks that in
reflection on history (understood as an academic discipline) it is
useful to adopt the following definition of theory:

> A theory, in the most elementary sense, is an effort to specify
> the relations among concepts in a law - like proposition *without
> regard to space or time.* The propositions are deductively con-
> nected and are, depending on their relationship with one
> another, either axioms (generalizations whose truth is taken
> for granted) or theorems (which are deduced from the
> axioms).(3)

This definition is, of course, used as a substantiation of the
statement that history lacks theory. In Hollingsworth's opinion,
this is due to improper research procedures used by historians,
and in particular to the historians' aversion to universal 'measurable' concepts:

> As almost every book in the philosophy of science points out,
> concept formation is one of the first and most important steps
> in any effort to construct a theory. . . . For most physical
> and biological scientists, the suggestion that concepts must not
> be culture bound is almost too obvious to mention. Even so,
> there is strong opposition to this view among most social
> scientists. And it is this resistance which has, as much as
> anything, hampered the development of theory in the social
> sciences.(4)

Hollingsworth also thinks that 'for purposes of theory construction, a concept should be measurable as a variable',(5) in
which connection he claims that from the point of view of theory
construction 'dichotomous' concepts are useless. And here are
examples of such useless concepts: Gemeinschaft and Gesellschaft
(Tönnies), mechanical and organic solidarity (Durkheim), folk
and urban societies (Robert Redfield), primary and secondary
relationship (Cooley), traditional and rational sources of
authority (Max Weber), inner directed and other directed
individuals (David Riesman).(6)

In Hollingsworth's text we can, of course, find many other
diagnoses, recommendations and items of practical advice ('. . .
one begins theory construction on a small scale, preferably with

the linking of two variables',(7) which resemble diagnoses, recommendations and practical advice discussed in chapter 5. Sociologist Erik Allardt, when stating that Hollingsworth's standpoint 'is quite clearly positivistic', comments pertinently that 'in the '50s and '60s most sociologists presenting views on theory-building would have offered views resembling those developed by Rogers Hollingsworth.'(8)

There is much evidence that sociology functions as the standard discipline with respect to other, even more 'retarded', social sciences. Are we to conclude from this, that the methodological consciousness shaped in the early twentieth century will - owing to the excellent articulation it has been given by the philosophy of science - persevere unchanged until the end of the century and continue to shape the profile of the social sciences the whole time?

Notes

1 INTRODUCTION

1 Paul K. Feyerabend, On the Critique of Scientific Reason, in R.S. Cohen et al. (eds), 'Essays in Memory of Imre Lakatos', Reidel, Dordrecht, 1978, p. 137.
2 I deliberately do not mention the Lvov-Warsaw School, despite its sometimes being classed as logical-positivist. For all the theoretical and technical similarities between that school and the Vienna Circle, representatives of the Lvov-Warsaw School were on the whole neutral with respect to the methodological doctrine of logical positivism. It is worth mentioning in this connection, that Stanisław Ossowski came from that school, which did not prevent him from being one of the most consistent and effective critics of the logical positivist doctrine when applied to the social sciences. I do realize, however, that the above thesis is disputable.
3 Cf. Pierre Bourdieu, The Specificity of the Scientific Field and the Social Conditions of the Progress of Reason, 'Social Science Information' 14(b). Bourdieu treats a scientific discipline 'as the locus of a political struggle for scientific domination' (p. 22), which makes him formulate very radical but excellently substantiated statements on science. It is, however, characteristic that he mainly refers to data pertaining to sociology, which is probably due not only to the tendency to make use of examples from one's own field. Moreover, whenever he reaches for examples from natural science, his strong statements resemble a set of idealizational propositions which explains a given situation well, but, as it were, sharpens its features; when he refers to examples drawn from sociology, the reader has the impression that his claims are in full agreement with facts.
4 Charles Ellwood, 'Methods in Sociology', Duke University Press, Durham, North Carolina, 1933; Pitirim Sorokin, 'Fads and Foibles in Modern Sociology and Related Sciences', Regnery, Chicago, 1956; C. Wright Mills, 'The Sociological Imagination', Oxford University Press, New York, 1959; Stanisław Ossowski, 'O osobliwościach nauk społecznych', (On the Peculiarities of the Social Sciences), PWN, Warsaw, 1962.

Philosophers of science may be interested in the fact that in Ellwood's book we find the following statement:

The history of all sciences shows that science develops, not through lingering in the field of factual description, not through timidity and caution, but through the development of bold hypotheses, and then by testing these hypotheses, not only through further research, but through actual experience and experimental action. Factual science is dead science, and should not be tolerated in institutions of learning which accept any degrees of responsibility for leadership in our civilization (p. 9).

First of all, we see here - one year before the appearance of 'Logik der Forschung' - the idea upon which Popper based his methodological model of the development of science. I do not think that this should induce us to draw far-reaching conclusions. In any case we are not to conclude that Ellwood's methodology was similar to Popper's. The passage above shows that Ellwood would develop that general idea quite differently than Popper did: for instance, with Ellwood, formulating bold hypotheses in order to test them is characteristic of procedure in science, and not of procedures used by the various scientists; Popper's methodological individualism inclined him to propagate a very naive picture of the scientists and the community of scientists. Nevertheless, the idea of conjectures and refutations is to be found in a book written by a sociologist who was only marginally concerned with methodological problems. I do not think that this fact should be explained by the movement of ideas - in either direction. It seems rather as though Popper constructed his system on the basis of an idea which, when compared with the doctrine of inductionism, was an intellectual event, but which is not particularly revealing for people doing research themselves.

5 F.A. Hayek, 'The Counter Revolution of Science', Free Press, Chicago, 1952.
6 It is to be noted that in English the term humanistic sociology is not very well defined and often occurs in trivial and pretentious context. We have to agree with Dennis Wrong that ' . . . "humanistic" has acquired a certain self-congratulatory aura, a "more-caring-about-people-than-thou" flavor.' (D. Wrong, 'Skeptical Sociology', Columbia University Press, New York, 1976, p. 2.)
7 Stanisław Ossowski, 'Dzieła' (Collected Works), vol. IV, PWN, Warsaw, 1967, p. 252.
8 Karl R. Popper, 'Objective Knowledge', Oxford University Press, 1974, pp. 185ff.
9 Ibid., p. 185.
10 Cf. Imre Lakatos 'The Mathematics of Collective Action', Heinemann Educational Books, London, 1973.
11 Preface to Anthony Giddens (ed.), 'Positivism and Sociology', Heinemann, London, 1974, p. ix.
12 Christopher G. A. Bryant, Positivism Reconsidered, 'The Sociological Review', vol. 23, no. 2, May 1975, p. 405.

13 Cf. A. Oberschall (ed.), 'The Establishment of Empirical
    Sociology', Harper & Row, New York, 1972.
14 Robert A. Dahl, Introduction to N. W. Polsby, R.A. Dentler
    and P.S. Smith (eds), 'Politics and Social Life: An Introductic
    to Political Behaviour', Houghton Mifflin, Boston, 1963.
    Quoted from: Michael Polanyi, 'Knowing and Being', Univer-
    sity of Chicago Press, 1974, p. 29.
15 Abraham Kaplan, Logical Empiricism and Value Judgements,
    in Paul A. Schilpp (ed.), 'The Philosophy of Rudolf Carnap',
    Open Court, La Salle, Illinois, 1963, p. 827.
16 I think that this is the basis of the opposition between
    empiricist and humanistic sociology in Stefan Nowak. The
    empiricist approach is described by reference to Lundberg's
    and Neurath's opinions on the subject matter and the con-
    ceptual apparatus of sociology. The approach typical of
    humanistic sociology is illustrated by statements by Znaniecki
    Max Weber, and MacIver, but is constructed as the antithesis
    of social behaviorism and physicalism. The program of
    humanistic sociology is thus reduced to reserving a proper
    place for the sphere of consciousness in sociological research,
    and understanding (in the sense of *Verstehen*) is reduced to
    operations defined by their goal, which is to ascribe mental
    states to human beings. Cf. Stefan Nowak, 'Studia z
    metodologii nauk społecznych' ('Studies in the Methodology of
    the Social Sciences'), PWN, Warsaw, 1965, chapter on
    Observation and understanding of human behavior versus
    problems of theory construction.
17 I wrote briefly about Polish sociology from that point of view
    in From Social Knowledge to Social Research: The Case of
    Polish Sociology, 'Acta Sociologica', vol. 17, no. 1, 1974.

2 THE HISTORY-ORIENTED TREND IN THE PHILOSOPHY OF SCIENCE: BREAKTHROUGH OR CONTINUATION?

1 This description of the effects of the crisis requires, however
   at least one addition: a rapid turnover of new methodological
   conceptions which in most cases were attempts so to modify
   Popper's doctrine as to make it withstand confrontation with
   historical data.
2 Ernam McMullin, The History and Philosophy of Science:
   A Taxonomy, in 'Minnesota Studies in the Philosophy of
   Science', vol. V, University of Minnesota Press, 1970, p. 18.
3 Imre Lakatos, History of Science and Its Rational Reconstruction, in 'Boston Studies in the Philosophy of Science', vol.
   VIII.
4 Ibid., p. 107.
5 Ibid., p. 93 (italics in the original).
6 Ibid., p. 106.
7 Ibid., p. 107. The appraisal of the sociology of science (or
   parts of it?) is equally severe: 'The work of those "external-

ists" (mostly trendy "sociologists of science") who claim to do
social history of some scientific discipline without having
mastered the discipline itself and its internal history, is
worthless.' (Ibid., footnote 68, p. 128.)
8 Cf. Thomas S. Kuhn, Notes on Lakatos, in 'Boston Studies
in the Philosophy of Science', vol. VIII, p. 141.
9 Imre Lakatos, op. cit., pp. 106-7.
10 Imre Lakatos, Falsification and the Methodology of Scientific
Research Programmes, in 'Criticism and the Growth of
Knowledge' (eds), I. Lakatos and A. Musgrave, Cambridge
University Press, 1970, p. 138. McMullin writes on that in
Ernam McMullin, op. cit., p. 33.
11 Opinions similar to those of Lakatos have been advanced by
Agassi:

> The history of science is a most rational and fascinating
> story; yet the study of the history of science is in a lament-
> able state: the literature of the field is often pseudo-
> scholarly and largely unreadable. The faults which have
> given rise to this situation, I shall argue, stem from the
> uncritical acceptance, on the part of historians of science,
> of two incorrect philosophies of science. These are, on the
> one hand, the *inductive philosophy* of science, according to
> which scientific theories emerge from facts, and on the
> other hand, the *conventionalist philosophy* of science,
> according to which scientific theories are mathematical
> pigeonholes for classifying facts. The second, although
> some improvement on the first, remains unsatisfactory.
> A third, contemporary theory of science, Popper's *critical
> philosophy* of science, provides a possible remedy. (J.
> Agassi, 'Toward an Historiography of Science', Mouton,
> 1963, introductory note.)

> Using that method of restoring the history of science to
> health Agassi wrote the book 'Faraday as a Natural Philoso-
> opher' (Chicago University Press, 1972). Other books which
> use an analogous method of pursuing history of science
> include 'Fields of Force, William Berkson (Routledge & Kegan
> Paul, 1974). Those two books provoked L. Pearce Williams,
> author of a comprehensive biography of Faraday, to write the
> paper Should Philosophers be Allowed to Write History?, in
> which he pointed to serious technical shortcomings of both
> authors and to evident distortions of facts, which were
> largely due to twisting historical data so that they should
> comply with Popper's methodological conceptions. Williams
> wrote in conclusion: 'I mean the question [see the title of
> the paper] quite seriously and I shall now answer it with a
> resounding "No!".' ('The British Journal for the Philosophy
> of Science', vol. 26, no. 3, September 1975, p. 252.)

12   We find, then, that there is not a single rule, however

plausible, and however firmly grounded in epistemology,
that is not violated at some time or other. It becomes eviden
that such violations are not accidental events, they are not
the results of insufficient knowledge or of inattention
which might have been avoided. On the contrary, we see
that they are necessary for progress. Indeed, one of the
most striking features of recent discussion in the history
and philosophy of science is the realization that developments such as the Copernican Revolution, or the rise of
atomism in antiquity and recently (kinetic theory; disper-
sion theory; stereochemistry; quantum theory), or the
gradual emergence of the wave theory of light occurred
either because some thinkers decided not to be bound by
certain 'obvious' methodological rules or because they
unwittingly broke them. (Against Method: Outline of an
Anarchistic Theory of Knowledge, 'Minnesota Studies in the
Philosophy of Science', vol. IV, p. 22.)

13   A methodology that emerges from an analysis of this kind
[i.e., a historical analysis, and in particular historical
case studies] differs from the existent system by its lack
of dogmatism and by its openness. Each rule, each demand,
that it contains is asserted only conditionally, like a rule
of thumb, and it can be overthrown, or replaced by its
opposite, as a result of an examination of concrete cases
(this applies even to such 'fundamental' demands as the
demand for consistency, for falsification, for agreement
with observational results, for maximal content in given
conditions, and so on). It is not claimed that the rules or
prescriptions are now derived from facts (though there
would not be anything wrong even with the procedure).
It is only asserted that inspection of facts educates a
sensible person and makes him aware of demands whose
existence, urgency and relevance he had not realized
before. Nor is it possible to say in advance under what
circumstances a rule may be suspended and how one will
behave then. The future cannot be foreseen and our actions
under new and yet untried circumstances cannot be pre-
dicted (after all, we do learn, and we do have new ideas).
Also, looking back into history, we find that for every
rule one might want to defend, there exist circumstances
where progress was made by breaking the rule. All this
means that methodology can at most offer a somewhat
chaotic list of rules of thumb and that the only principle
we can trust under all circumstances is that *anything goes*.
(Problems of Empiricism, Part II in 'The Nature and
Function of Scientific Theories' (ed.), Robert G. Colodny,
University of Pittsburgh Press, 1970, pp. 277-87).

14 Ibid., p. 277.
15 Case studies are to be found in various research contexts,

and hence they do not make any category of methodologically
uniform operations. Among other things they serve as a
method of explication, based on concrete material, of a theory,
conception, etc. Now Feyerabend's case studies are clearly
in the last-named category. Their objective, which is easy to
decipher, is to explain a certain methodological conception of
science, adopted in advance and shaped following philosophical
discussions. In Feyerabend's works, history plays a heuristic
role: it helps him present philosophical conceptions and argue
in their favor, but the source and the real grounding of these
conceptions is to be sought elsewhere, and the relation they
bear to historical facts remains an open issue.

16 E. McMullin, op. cit., p. 41.
17 Cf. Alvin W. Gouldner, 'The Coming Crisis of Western
   Sociology', Heinemann, London, 1971.
18 When comparing his views with those of Popper, Kuhn writes:

> On almost all occasions when we turn explicitly to the same
> problems, Sir Karl's view of science and my own are very
> nearly identical. We are both concerned with the dynamic
> process by which scientific knowledge is acquired rather
> than with the logical structure of the products of scientific
> research. Given that concern, both of us emphasise, as
> legitimate data, the facts and also the spirit of actual
> scientific life, and both of us turn often to history to find
> them. From this pool of shared data, we draw many of the
> same conclusions. Both of us reject the view that science
> progresses by accretion; both emphasise instead the revolu-
> tionary process by which the older theory is rejected and
> replaced by an incompatible new one; and both deeply
> underscore the role played in this process by the older
> theory's occasional failure to meet challenges posed by
> logic, experiment, or observation. Finally, Sir Karl and I
> are united in opposition to a number of classical positivism's
> most characteristic theses. We both emphasise, for example,
> the intimate and inevitable entanglement of scientific obser-
> vation with scientific theory; we are corresondingly
> sceptical of efforts to produce any neutral observation
> language; and we both insist that scientists may properly
> aim to invent theories that *explain* observed phenomena
> and that do so in terms of real objects, whatever the latter
> phrase may mean.
>   That list, though it by no means exhausts the issues
> about which Sir Karl and I agree, is already extensive
> enough to place us in the same minority among contemporary
> philosophers of science. (Logic of Discovery or Psychology
> of Research?, in 'Criticism and the Growth of Knowledge',
> op. cit., pp. 1-2.)

Some lines further Kuhn states: 'My agreement with Sir
Karl is real and substantial.' Those declarations are not to be

taken at their face value: one of their roles is to be a politely
worded introduction to the discussion of the differences
between Kuhn's and Popper's views, which differences are
described by Kuhn as a 'gestalt switch rather than a disagreement'. Popper in his reply ignores the 'gestalt switch'
stressed by Kuhn, and emphasises the differences of opinion
between them on certain important issues. (Cf. K. Popper,
'Normal Science and its Dangers', in the same book.)

19 Cf. Stefan Amsterdamski, afterword to the Polish-language
edition of Kuhn's 'Structure of Scientific Revolutions', PWN,
1968; 'Between Experiment and Metaphysics', Książka i
Wiedza, 1973; Science as a Subject Matter of Humanistic
Reflection, 'Studia Socjologiczne', 1971, no. 2 (41) (all items
in Polish).

20 History of Science and its Rational Reconstruction, op. cit.,
p. 115.

21 Thomas S. Kuhn, 'The Structure of Scientific Revolutions',
University of Chicago Press, 2nd edn, p. 9 (author's italics).

22 Karl Popper, 'The Logic of Scientific Discovery', Harper
Torchbooks, Harper & Row, New York and Evanston, 1968,
p. 35.

23 Ibid., p. 37.

24 I. Lakatos, History of Science and its Rational Reconstruction
op. cit., p. 110.

25 One of the points is that Popper is not versed in social
science, and his appraisals in that field are usually unsub-
stantiated or based on non-professional interpretations. Note,
for instance, Popper's attitude toward the Frankfurt school.
When referring to his discussion with Theodor Adorno, Popper
wrote:

> Having been invited to speak about 'the Logic of Social
> Science' I did not go out of my way to attack Adorno and
> the 'dialectical' school of Frankfurt (Adorno, Horkheimer,
> et al.), which I never regarded as important, unless per-
> haps from a political point of view; and in 1960 I was not
> even aware of the political influence of this school. Although
> today I should not hesitate to describe this influence by
> such terms as 'irrationalist' and 'intelligence-destroying',
> I could never take their methodology (whatever that may
> mean) seriously from either an intellectual or a scholarly
> point of view. (K.R. Popper, Reason or Revolution?, in
> 'The Positivist Dispute in German Sociology', Heinemann,
> London, 1976, p. 289.)

In the same text Habermas is characterized thus: 'Most of
what he says seems to me trivial; the rest seems to me mistaken.' (Ibid., p. 297.)
The point is that Popper leaves those opinions unsub-
stantiated. It is unthinkable that he would treat any significant school in physics in the same way. In my opinion he

would not deal thus with any philosophical school, any school of painting, any poetic group or any religious sect. In Popper's opinion, the production of the school of Frankfurt should, unlike philosophy or poetry, meet the criteria of what science is as he interprets them, and even a cursory analysis shows that it does not meet them. He accordingly believed that he could confine himself to such a cursory analysis. I claim that this is the source of all those opinions of Popper's on social science which do not immediately follow from the assumption that the methodological characteristic of social science consists of the characteristic of natural science plus the thesis on retardation.

26 Cf. Reflections on my Critics, in 'Criticism and the Growth of Knowledge', (eds) I. Lakatos and A. Musgrave, Cambridge University Press, 1970.

27 Ibid., p. 244.

28 Ibid., p. 245.

29 Jürgen Habermas, 'Erkenntnis und Interesse', Suhrkamp Verlag, 1968.

30 J.D. Bernal begins his 'Science in History' with the palaeo-lithic period. Research on Chinese science holds, mainly owing to Joseph Needham and his school, a very important place in Western historiography. Anthropological case studies concerned with pre-literate peoples often include sections entitled 'science' or 'magic and science'.

3 THE PHILOSOPHY OF SCIENCE IN THE PERSPECTIVE OF THE THEORY OF CULTURE

1 Imre Lakatos, History of Science and its Rational Reconstruction, 'Boston Studies in the Philosophy of Science', vol. viii, p. 107.

2 Bronisław Malinowski, 'Magic, Science and Religion', Free Press, Chicago, Illinois, 1948, p. 1.

3 Ibid.

4 In another place Malinowski seems even to identify science with 'real', 'true', knowledge: 'From the very beginnings man must have acquired *real, that is, truly scientific,* knowledge in order to transmit the earliest technological inventions.' ('Freedom and Civilization', Roy Publishers, New York, 1944, p. 206. Author's italics.)

5 'Magic Science and Religion', op. cit., p. 17.

6 Ibid., p. 13.

7 Judith Willer, 'The Social Determination of Knowledge', Prentice-Hall, Englewood Cliffs, New Jersey, 1971, p. 11.

8 Ibid., p. 25.

9 Ibid., p. 31.

10 Ibid., p. 10. Miss Willer's conclusions pertain to the non-scientific nature of the various academic disciplines. 'So-called (political) theory is no more than a simple empirical general-

ization;' 'most of economic theory is likewise empirical;' 'the anthropologist, too, whatever his claim, is typically an empirical thinker utilizing a magical system of knowledge just like his more primitive subjects;' 'sociology also claims to be scientific, but like anthropology, it lacks theory.' (Ibid., pp. 142-3.)

11 Florian Znaniecki, 'The Social Role of the Man of Knowledge', Octagon Books, New York, 1965, p. 198.

12 Ibid., p. 192.

13 Ibid., p. 11.

14 Ibid., p. 12.

15 Interpretation of science in terms of social roles and institutions has become extremely popular in sociological literature thanks mainly to Znaniecki's influence. For a radical exposition of that standpoint see David Bloor, 'Knowledge and Social Imagery', Routledge & Kegan Paul, London, 1967.

16 A good criticism of Malinowski's theoretical conceptions is to be found in E.R. Leach, The Epistemological Background to Malinowski's Empiricism, in 'Man and Culture. An Evaluation of the Work of Bronisław Malinowski, Raymond Firth (ed.), first published by Routledge & Kegan Paul in 1957. Cf. other items in that book. See also: Bryan R. Wilson (ed.), 'Rationality', Harper & Row, New York, Evanston, San Francisco, London, 1971. Robin Horton and Ruth Finnegan (eds), 'Models of Thought', Faber & Faber, London 1973.

17 Moritz Schlick, 'General Theory of Knowledge', Springer-Verlag, New York and Vienna 1974, p. 4.

18 In other words, in accordance with the present-day interpretation, this requirement applies not to the research procedure, but to its products.

19 This is not to say that these concepts are determined by those assumptions. Marx's concept of social class is clearly limited by such elements of his system as the holistic approach to social groups which greatly restricts the possibilities of interpretation (e.g., it makes one stress the difference between social class and socio-economic category). The assumptions of Marx's system assign a (very high) rank to that concept, the limits of its transformations within the system (e.g., following the emergence of new social conditions), etc. (There is, of course, always the problem, which theses are to be accepted as the basic assumptions of the system and how to interpret them, but that is another, although related, issue.) Yet no set of assumptions in Marx's system makes it possible to grasp that concept, to explain it, or to use it. It may thus be said that the assumptions of the system provide certain criteria of applicability of the term 'social class', but those criteria do not enable one to use the term correctly; such possibilities are offered only by an analysis of the system as a whole.

20 Jerzy Kmita, How to Cultivate the Methodology of Sciences (in Polish), Studia Filozoficzne, no. 1, 1975, p. 82.

1 Hans Reichenbach, 'Experience and Prediction: An Analysis of the Foundations and the Structure of Knowledge', Phoenix Books, University of Chicago Press, 1961, p. 3.

2 Ibid., pp. 5-6. ('Rational reconstruction' is italicized in the original, the others are author's italics.)

3 Ibid., p. 6. (Author's italics.)

4 Cf. W.V. Quine, Epistemology Naturalized, in Roderick M. Chisholm and Robert J. Swartz (eds), 'Empirical Knowledge: Readings from Contemporary Sources', Prentice-Hall, Englewood Cliffs, New Jersey, 1973.

5   Around 1932 there was debate in the Vienna Circle over what to count as observation sentences, or protokollsätze. One position was that they had the form of reports of sense impressions. Another was that they were statements of an elementary sort about the external world, e.g., 'A red cube is standing on the table.' Another, Neurath's, was that they had the form of reports of relations between percipients and external things: 'Otto now sees a red cube on the table.' The worst of it was that *there seemed to be no objective way of settling the matter; no way of making a real sense of the question.* (W.V. Quine, op. cit., author's italics.)

But was it not possible to *verify empirically* what plays the role of observation statements in science and whether anything plays such a role? The problem, however, was differently formulated. It was known, on the basis of the epistemological doctrine adopted, that knowledge derives from direct empirical data; it was known, accordingly, that statements presenting those data do function in science and play a fundamental role in it. The problem to be settled was: what form should a statement have to express direct empirical data? It was thus, in fact, the question: what are direct empirical data?, and hence an inner problem of the epistemological doctrine of empiricism.

6   P.K. Feyerabend, On the Critique of Scientific Reason, in R.S. Cohen et al. (eds), 'Essays in Memory of Imre Lakatos', D. Reidel, Dordrecht, 1976, p. 113.
This applies primarily to the trend connected with the legacy of the Vienna Circle. Popper's philosophy of science was from its very inception more linked to real science, which was a sui generis paradox if we consider the fact that the conception of the philosophy of science he had adopted left him a wider margin of freedom in that respect than the conception advocated by the Vienna Circle did. This does not alter the fact that 'the problems of science' analyzed in his works remain essentially inner problems of his methodological system. As has been said, Popper focused his analysis on 'Great Science', which included all and only that which in his intellectual circles enjoyed the status of great science (which is to say

that the scope of that Great Science was determined societal
His methodological model was, by assumption, to characteriz
the production of great personalities in science, 'such as
Galileo, Kepler, Newton, Einstein and Bohr (to confine myse
to a few of the dead).' He applied to them his thesis, which
he treated as a simplified empirical description. 'These are
men of bold ideas, but highly critical of their own ideas; the
try to find whether their ideas are right by trying first to
find whether they are not wrong. They work with bold conjecture and severe attempts at refuting their own conjecture
This concludes the empirical part. 'My criterion of demarcati
between science and nonscience is a *simple logical analysis o
this picture.*' (Karl Popper, Replies to My Critics, in Paul A
Schilpp (ed.), 'The Philosophy of Karl Popper', Book II,
Open Courts, La Salle, Illinois, 1974, pp. 977-8. Author's
italics.)

It is worth adding that the criterion of demarcation with
Popper, and in other authors, plays the role of the definitio
of science. In Popper's case this is done consciously: 'If I
define "science" by my criterion of demarcation (I admit tha
this is more or less what I am doing) . . . ' (Ibid., p. 981.

28 Cf. P.K. Feyerabend, Against Method: Outline of an
Anarchistic Theory of Knowledge, in 'Minnesota Studies in
the Philosophy of Science', vol. IV, p. 22; Problems of
Empiricism, Part II, in R. Colodny (ed.), 'The Nature and
Function of Scientific Theories', University of Pittsburgh
Press, 1970.

29 P.K. Feyerabend, On the Critique of Scientific Reason, op.
cit., p. 136.

30 In my opinion, this is the main explanation of Popper's shar
reaction to Kuhn's statement on the need for psycho-socio-
logical research on science:

> To me the idea of turning for enlightenment concerning th
> aims of science, and its possible progress, to sociology or
> to psychology (or, as Pearce Williams recommends, to the
> history of science) is surprising and disappointing.
>   In fact, compared with physics, sociology and psycholo
> are riddled with fashions, and with uncontrolled dogmas.
> The suggestion that we can find anything here like 'objec
> tive, pure description' is clearly mistaken. Besides, how
> can the regress to these often spurious science help us in
> this particular difficulty? Is it not sociological (or psycho
> logical, or historical) *science* to which you want to appeal
> in order to decide what amount to the question 'What is
> *science*?' or 'What is, in fact, normal in science?' For clea
> you do not want to appeal to the sociological (or psycho-
> logical or historical) lunatic fringe? (Normal Science and I
> Dangers, in I. Lakatos and A. Musgrave (eds), 'Criticism
> and the Growth of Knowledge', Cambridge University Pre
> 1970, pp. 57-8.)

This looks like the advice that lost objects should always
be looked for near a street lamp, because it is light there,
but the point in fact is that the very idea of empirical
research on science - by whatever methods - is disastrous
for the belief that 'the method of science is the method of
bold conjectures and ingenious and severe attempts to refute
them,' on which belief Popper's methodological system is based.
Equally disastrous for that system would be to treat it as a
strictly normative one. Its survival would then require
demonstration of its merits other than the alleged connections
with the research practice of Great Science. Popper must
maintain the unsteady balance between the postulate of correspondence and that of empirical uncontrollability of
methodological constructions. That balance is disturbed by
tendencies to engage in historical research on science and
that is what the significance of these tendencies consists of.

THE UNITY OF SCIENCE AS A 'LOGICAL NECESSITY'

'Empirische Soziologie. Der wissenschaftliche Gehalt der
Geschichte und Nationalökonomie', Julius Springer, Wien 1931.
Quoted from the English-language version, Empirical Sociology.
The Scientific Content of History and Political Economy, in
Otto Neurath, 'Empiricism and Sociology', M. Neurath and
R.S. Cohen (eds), D. Reidel, Dordrecht, 1973, p. 360.
Ibid., p. 325.
Of course, this assumption was not formulated explicitly.
Moreover, formulating it in such a language would be
inadmissible.
Quoted from the English-language version, The Logical
Analysis of Psychology, in Herbert Feigl and Wilfrid Sellars
(eds), 'Reading in Philosophical Analysis', Appleton-Century-
Crofts, New York, 1949, pp. 373-4.
As can easily be seen, Hempel presents his 'new analysis'
by means of such encouraging phrases as 'rigorous logical
tools', 'extremely subtle logical apparatus', and 'modern
logistic'. Paying attention to this can - and certainly will to
people emotionally connected with the tradition of logical
positivism - seem to be sheer pedantry. But the point is,
that using such phrases has largely contributed to the success
of logical positivists in the social sciences, a success which
exceeded their own expectations and at one moment became
a cumbersome ballast of their doctrine. Generally speaking,
that success was a phenomenon that can be explained in
social terms only, i.e., in terms of the social mechanisms of
the otherwise incomprehensible outburst of the popularity
of certain ideas. One of those mechanisms, not to be belittled
in spite of appearances, is connected with the role of symbols
of group status, in this case the symbols of what is scientific.
Logical positivism entered the social sciences from the position

of science, and not as an epistemological doctrine or a
methodological program. This was, in any case, the situation
as the recipients saw it. In consolidating that position the
language of logic, and especially logical symbolism (both the
fact that it was used and the emphasis placed on that fact)
played an enormous role. 'Rigorous and subtle logical
apparatus', regardless of how far it was really rigorous and
subtle, was accepted as a token of the scientific approach,
and among the social scientists it played the role of glass
beads in trade with primitive tribes: it was both the desired
symbol of being a member of the elite and, in many cases,
the only justification of that membership.

5   Ibid. p. 374.
6   Ibid.
7   Empirical Sociology . . . op. cit., p. 357.
8   R.J. Ackerman, 'The Philosophy of Karl Popper', University
    of Massachusetts Press, Amherst, 1976, p. 163. See also
    Alan Donagan, Popper's Examination of Historicism, in P.A.
    Schilpp (ed), 'The Philosophy of Karl Popper', Open Court,
    La Salle, 1974, Book II.
9   This is a very important element of the methodological doctr
    of logical positivism. Replacement of this seemingly self-
    evident thesis by another seemingly self-evident thesis,
    namely: 'The theoretical content of science is to be found in
    theories', could essentially change the history of the
    philosophy of science. The first thesis takes the proposition
    as the basic methodological category, and it was so inter-
    preted within logical-positivist doctrine. The second takes
    the theoretical system as the basic methodological category.
    In the former case, a scientific theory is a set of proposition
    that satisfy definite conditions, in the latter, a proposition
    is an element of a theory that satisfies definite conditions. I
    other words, in the former case the properties of good scien
    are determined by the properties of propositions, in the
    latter, by the properties of theories. The difference consist
    in the fact that, among other things, in the latter case, e.g
    verifiability is a property of a theoretical system and need
    not be an attribute of its various parts - unless we adopt
    additional assumptions which are far from self-evident. This
    difference is essential for the social sciences: if we assume
    that verifiability, empirical meaning, etc., are properties of
    theories, then nothing follows from the objections, stereo-
    typically formulated over the last fifty years, that propositi
    of classical sociological theories are unverifiable or lack
    empirical meaning. Likewise, if the theory is treated as the
    basic methodological category, this dismisses the requiremen
    that the meaning of a term be determined by a definite
    system of propositions, and thus puts an end to the basis o
    the logical-positivist criticism of the conceptual apparatus o
    the social sciences. This is not to say, of course, that start
    ing from a holistic vision of science one cannot formulate

criteria of scientific correctness which would be as unfavour-
able for the social sciences as the classical logical-positivist
criteria based on the atomistic vision of science.

It is worth noting that studies by such authors as Hanson,
Kuhn and Feverabend imply a holistic conception of scientific
system. We are, however, not in a position to analyze the
consequences of that approach for the traditional methodological doctrines, and especially for Carnap's doctrine, which
would be the most threatened in such a context.

10 C. Hempel, ibid., p. 377.
11 Ibid., p. 376.
12 Ibid., p. 378.
13 A clear case of a simple generalization. The statements given
by Hempel are not, in this text, examples which illustrate
the thesis, which has been otherwise accepted as true, but
form the inductive basis of the generalization. Hempel
obviously treats them as a sufficiently good representation of
all the statements that describe the conditions of verification
of the statement 'Paul has a toothache.' This is a quite
arbitrary decision. Note that this list of statements could be
extended infinitely (and not just 'much', as Hempel has it),
and Hempel's generalization concerning their nature must be
based on theoretical considerations, and not on an analysis of
several or even several thousand cases.
14 Ibid., pp. 377-8.
15 Ibid., p. 381. (Author's italics.)
16 This is how Carnap in his autobiography describes the origin
and evolution of physicalism:

In our discussions, chiefly under the influence of Neurath,
the principle of the unity of science became one of the
main tenets of our general philosophical conception. This
principle says that the different branches of empirical
science are separated only for the practical reason of
division of labor, but are fundamentally merely parts of
one comprehensive unified science. This thesis must be
understood primarily as a rejection of the prevailing view
in German contemporary philosophy that there is a fundamental difference between the natural sciences and the
Geisteswissenschaften. . . . In contrast to this customary
view, Neurath maintained the monistic conception that
everything that occurs is part of nature, i.e., of the
physical world. I proposed to make this thesis more precise
by transforming it into a thesis concerning language,
namely, the thesis that the total language encompassing all
knowledge can be constructed on a physicalistic basis. . . .
Our initial formulations of physicalism . . . can only be
regarded as a first rough attempt. In view of the liberalization of the empiricist conception which was achieved some
years later, the assertion of the definability of psychological terms on the basis of the terms of the thing-language

must be weakened to an assertion of reducibility. (R.
Carnap, Intellectual Autobiography, in: Paul A. Schilpp
(ed.), 'The Philosophy of Rudolf Carnap', Open Court, La
Salle, 1963, pp. 52-3.)

17 In the 1960s the custom still prevailed of dedicating one or
two last chapters in publications concerned with systematizing
the methodological doctrines, to the social sciences (*The
Structure of Science* by E. Nagel has three such chapters).
In the latest publications of this kind, written by the younger
generation of philosophers of science, this custom has practically fallen into disuse – to the advantage of the publication
involved incidentally. As the importance of methodology in the
social sciences increased, translation of the doctrine into the
language of the 'methodology of the social sciences' fell to
the sociologists (and also representatives of other social
disciplines) who specialize in methodological research. Texts
in that category are as a rule at the popular level (Cf. 'On
Theory and Verification in Sociology' by Hans Zetterberg).
In this connection 'The Conduct of Inquiry' by Abraham
Kaplan (Chandler, San Francisco, 1964) occupies an exceptional position, being – in my opinion – decidedly the best
and the most competent (when it comes to knowledge of both
the philosophy of science and the social sciences) publication
on the methodological foundations of the social sciences within
the framework of the philosophical tradition here under consideration.

18 Ibid., p. 382.

19 Ibid. Italics in the original.

20   We have witnessed a great many reductions of science to
other sciences (astronomy, acoustics, and thermodynamics
to mechanics, optics to electrodynamics) and such impressive
unifications as those of mechanics and electrodynamics as
brought about by the relativity and quantum theories. In
this connection it can be safely assumed today that
chemistry is becoming reducible to atomic physics.
Further unifications may be expected to emerge out of the
endeavours of such borderland sciences as bio-physics,
bio-chemistry, psycho-physics, psycho-physiology, social
psychology, etc. As to whether these syntheses will lead
to a complete reduction of the laws of the various sciences
to a unitary set of basic laws, and as to whether these
basic laws will be the laws of a future physics is again a
matter of conjecture. (H. Feigl, Logical Empiricism, in
H. Feigl and W. Sellars (eds), 'Readings in Philosophical
Analysis', Appleton-Century-Crofts, New York, 1949,
pp. 22-3).

21 H. Feigl, Unity of Science and Unitary Science, in H. Feigl
and M. Brodbeck (eds), 'Readings in the Philosophy of

Science', Appleton-Century-Crofts, New York, 1953, p. 384.
22 C. Hempel, ibid., p. 382.
23 R. Carnap, Logical Foundations of the Unity of Science, in
'International Encyklopedia of Unified Science', I, 1938;
quoted from the reprinted text in H. Feigl and W. Sellars
(eds), 'Readings in Philosophical Analysis', op. cit., p. 422.
That faith has never been extinguished, but it came to be
separated in practice from analyses proper by taking on the
form of descriptions of future conditions, descriptions that
resemble those socio-economic prognostic visions whose only
substantiation is the unfailing 'it will be so'. Thus, for
instance, when concluding his famous article on theoretical
terms, Carnap formulates the following prediction that con-
trasts in style with the importance of the text as a whole:

> Both these approaches in psychology (i.e., molar and
> theoretical) will probably later converge toward theories on
> the central nervous system formulated in psychological
> terms. In this physiological phase of psychology, which
> has already begun, a more and more prominent role will
> be given to quantitative concepts and laws referring to
> micro-states described in terms of cells, molecules, atoms,
> fields, etc. And finally, micro-physiology may be based on
> micro-physics. This possibility of constructing finally all
> of science, including psychology, on the basis of physics,
> so that all theoretical terms are definable by those of
> physics and all laws derivable from those of physics, is
> asserted by the thesis of physicalism (in its strong sense).
> (The Methodological Character of Theoretical Concepts, in
> 'Minnesota Studies in the Philosophy of Science', vol. I,
> University of Minnesota Press, Minneapolis, 1956, pp. 74-5.)

24 Neurath thought (cf. his 'Empirische Soziologie') that the
unity of science is of value in relation to the practical functions of science, which functions he linked with prediction,
in agreement with the concept adopted in logical positivism.
He noted that predicting definite events often involves the
necessity of availing oneself of knowledge in various fields:
physics, biology, psychology. From this he drew the conclusion that integration of science was necessary, without
explaining however, what would be the advantages of such
integration, i.e., why availing oneself of the knowledge
drawn from the various disciplines (not integrated in the
logical-positivist sense of the word) complicates matters.
25 'Erkenntnis', Vol. 2, No. 5/6, Vol. 3, No. 2/3, 1932.
26 'Philosophy of Science', 3/1936 and 4/1937.
27 'Wissenschaftliche Weltauffassung, Der Wiener Kreis',
Artur Wolf, Wien, 1929. Quoted from the English-language
version: Otto Neurath, 'Empiricism and Sociology', Marie
Neurath and Robert S. Cohen (eds), Reidel, Dordrecht,
1973, chapter 9, p. 309.

28 'The backward state of the Moral Sciences can only be
   remedied by applying to them the methods of Physical
   Science, duly extended and generalized.' (John Stuart Mill,
   'A System of Logik', contents, book VI, chapter I, Section 1.
29 M. Schlick, 'Empirical Sociology', op. cit., p. 355.
30 The Function of General Laws in History, 'The Journal of
   Philosophy', 39, 1942. Quoted from the reprinted version in:
   H. Feigl and W. Sellars, 'Readings in Philosophical Analysis',
   op. cit. See also the less known but very similar (both in
   the choice of problems and the arguments used) paper by
   Ernest Nagel, written ten years later, namely Some Issues in
   the Logic of Historical Analysis, in P. Gardiner (ed.),
   'Theories of History', Free Press, Chicago, 1959. Unlike
   Hempel, Nagel refers directly to Rickert's studies.
31 Ibid., p. 459.
32 Ibid., p. 467. (Author's italics.)
33 Ibid., p. 459.
34 That schema also occurs in the article under consideration
   (for the first time to my knowledge, even though the corresponding conception of explanation was presented in fairly
   great detail in Popper's 'Logik der Forschung'), but in a more
   descriptive form:

> The scientific explanation of the event in question consists
> of
> (1) a set of statements asserting the occurrence of certain
>   events $C_1, \ldots C_n$ at certain times and places,
> (2) a set of universal hypotheses, such that
>   (a) the statements of both groups are reasonably well
>   confirmed by empirical evidence,
>   (b) from the two groups of statements the sentence
>   asserting the occurrence of event E can be
>   logically deduced (p. 460).

   That schema was successively given more and more precision in The Logic of Explanation (1948), The Logic of
   Functional Analysis (1959), Deductive-Nomological vs
   Statistical Explanation (1962), and Aspects of Scientific
   Explanation (1965). These texts, except for the third one,
   are included in Carl G. Hempel, 'Aspects of Scientific Explanation', The Free Press, New York, 1965. Hempel did not
   make any essential changes, but expanded the problems of
   statistical explanation, barely mentioned in the said article.
35 Ibid., pp. 365-6.
36 Ibid., p. 468.
37 'Economic Behavior' by D.W. McConnel et al., (an outline
   of the institutional approach in economics).
38 Ibid., p. 464.
39 For more on this subject see my The Methodological Dogma
   of Naturalism, in 'Studies in Methodology', Polish Sociology
   1975-76, Ossolineum, 1977.

40 Karl R. Popper, 'The Poverty of Historicism', Basic Books, New York, 1960, p. 2.
41 Ibid., (author's italics).
42 As is known, Neurath was enthusiastic about Marxism and considered himself a Marxist. He also explained the situation in the philosophy of the social sciences 'in the Marxist way' by pointing to its connections with socio-political conditions. He claimed that up to this time behaviourism could have developed only in the United States and in the Soviet Union, whereas in Germany aversion to behaviourism was particularly strong, and ever naturalistic, i.e., scientific, sociology was rejected there. In more general terms, 'The struggle between metaphysical and scientific sociologists cannot be avoided, it reflects the much stronger social conflicts.' ('Empiricism and Sociology', op. cit., p. 361.) His Marxist leanings induced Neurath to give an exceptionally favourable assessment of that system from the point of view of physicalistic assumptions:

> The materialist conception of history may err on this or that point, but in any case it figures among those conceptions of history which in principle can transform everything they say into physicalist language, that is to say, into the language of 'sociology of materialist foundation'. (p. 358).

Neurath also formulated the supposition that the further evolution of Marxism

> will lead to a transformation of statements concerning the 'substructure' into statements concerning the energy turnover of nutrition, heating and so on, whereas a part of the statements concerning the 'superstructure' will try to assess the energy turnover in games, ceremonials, court proceedings and so on, and a part of the statements concerning the 'superstructure' might be set apart as statements about peculiarities of structure. (p. 353).

43 Cf. vol. I of 'Minnesota Studies in the Philosophy of Science', University of Minnesota Press, Minneapolis, 1956.
44 Cf. chapter 2 of the present book.
45 As has been noted by Christopher Bryant, these criteria were undergoing certain modifications, which in turn modified the answer to the question, in what the retardation of sociology consists. Cf. C.G.A. Bryant, Kuhn, Paradigms and Sociology, 'The British Journal of Sociology', vol. 26, no. 3, 1975.

5 FROM METHODOLOGICAL DOCTRINE TO RESEARCH PRACTICE

1 R. Carnap, The methodological character of theoretical concepts, in 'Minnesota Studies in the Philosophy of Science',

vol. I. University of Minnesota Press, Minneapolis, 1956, p. 70 (author's italics).

2 Mario Bunge, one of the more liberal philosophers of science postulates: A moral for philosophical critics of science: First bow, then strike. (M. Bunge, 'Scientific Research I: The Search for System', Springer-Verlag, New York, 1967, p. 19.) In the same chapter he attacks psychoanalysis (which he considers a pseudoscience like dowsing and parapsychology), and does it in a way which not only reveals a lack of defence which he postulates, but also evident ignorance of the subject.

3 B.F. Skinner, 'Science and Human Behavior', The Free Press, Chicago, 1953, p. 5.

4 Cf. G. Lundberg, 'Can Science Save Us?', Longmans, Green, New York, 1974.

5 Cf. A.R. Radcliffe-Brown, 'A Natural Science of Society', Free Press, Chicago, Ill., 1957.

6 Cf. Stuart C. Dodd, A Barometer of International Security, 'Public Opinion Quarterly', Summer 1945.

7 E.T. Bell, a review of 'Dimensions of Society' in 'The American Sociological Review', vol. vii (1942), no. 5.

8 Ibid., p. 709.

9 See T. Parsons's review in the same issue of 'The American Sociological Review'.

10 Cf. P. Bourdieu, The Specificity of the Scientific Field and Social Conditions of the Progress of Reason, 'Social Science Information', 14 (6).

11 As was noted by Richard J. Hill (On the Relevance of Methodology, in N.K. Denzin (ed.), 'Sociological Methods', Butterworths, London, 1970) specialization in American sociology has led to the disappearance of the category 'general methodologist' and the emergence of narrower specialization.

   Younger scholars are . . . experts on Markov processes or computer simulation or multi-dimensional measurement models or some other specialty which is viewed by many of their colleagues as esoteric scientism having little relevance for the general substantive problems of sociology (p. 14).

12 Cf. 'Analizy i próby technik badawczych' ('Analyses and Testing of Research Techniques') by J. Lutyński and Z. Gostkowski (eds), Ossolineum, Wrocław, vol. 1: 1966, vol. 2: 1968.

13 See in particular H. Blumer, 'An Appraisal of Thomas and Znaniecki's "The Polish Peasant in Europe and America," Critiques of Research in the Social Science, I', Social Science Research Council, New York, 1939; Ch. Cooley, 'Sociological Theory and Social Research', Henry Holt, New York, 1930; R. MacIver, 'Social Causation', Ginn, Boston, 1942; F. Znaniecki, 'The Method of Sociology', Farrar & Rinehart, New York, 1934; Ch. Ellwood, 'Methods in Sociology', Duke University Press, Durham, 1933; H. Becker, 'Sociological

Work: Method and Substance', Aldine, Chicago, 1970; A.
Cicourel, 'Method and Measurement in Sociology', Free Press,
New York, Ill., 1964.
14 The quotation marks are used here to stress the fact that we
mean the subdiscipline which functions under that name, and
not the methodology of social research in general. In the
light of this distinction there is no contradiction in Howard
Becker's statement that 'methodology is too important to be
left to methodologists' (p. 3 of his book quoted in footnote
(13)). The relation which that (historically shaped and
institutionally established) subdiscipline bears to the general
methodology of social research is an open issue. As has been
noted by Howard Becker,

the question . . . arises as to whether methodologists – the
institutionally accepted guardians of methodology – deal with
the full range of methodological questions relevant for
sociology, or whether they deal with non-randomly selected
subset (as they might say) of those questions (ibid.).

In my opinion, Becker made a too hasty assumption that the
problems with which the methodologists of social research
deal, are totally in the class of the problems of the methodology
of social research (without quotation marks).
15 Hubert Blalock reflected on solving the following dilemma: the
constant development of methods of quantitative studies is
increasing the pressure on the curricula in that field. On the
other hand, there is a tendency to reduce the curricula in
their obligatory parts. How is that problem to be solved?
One of the plausible solutions is: It is also possible to reduce
the number of course requirements in all areas except methodology. As it can be judged from the context, Blalock would
consider that a good solution were it not for problems of
organization:

"But this creates certain strains in the system, and places a
burden on methodology specialists to justify the addition of a
new requirement, or even the retention of the old ones. The
curriculum appears to be top-heavy with methodology, and
students will begin applying pressure to balance things out
by doing away with this one-sidedness (H.M. Blalock, Jr.,
On Graduate Methodology Training, 'The American Sociologist',
vol. 4, no. 1, February 1965, p. 5)

16 C. Selltiz et al., 'Research Methods in Social Relations',
revised one-volume edition, Holt, Rinehart & Winston, New
York, Chicago, San Francisco, Toronto, 1964, p. 2.
17 D. Willer, 'Scientific Sociology: Theory and Method', Prentice-
Hall, Englewood Cliffs, 'New Jersey', 1967, pp. xiv and xix
(author's italics).
18 E.F. Borgatta, The Current Status of Methodology in Sociology,

in E.F. Borgatta (ed.), 'Sociological Methodology 1969',
Jossey-Bass, San Francisco, 1969, p. x.
19 Cf. F. Adler, Comments on Lundberg's Sociological Theories,
in A. de Grazia et al. (eds), 'The Behavioral Sciences Essays
in Honor of George A. Lundberg', The Behavioral Research
Council, Great Barrington, Mass., 1968.
20 P.F. Lazarsfeld, Philosophy of Science and Empirical Social
Research, in E. Nagel et al. (eds), 'Logic, Methodology
and Philosophy of Science', Stanford, 1962.
21 Cf. H.L. Zetterberg, 'On Theory and Verification in Sociology'. In the first edition of that extremely influential and
popular book (Almqvist & Wiksells, Stockholm, and Tressler
Press, New York, 1954), its author mentions four publications
on logic and the philosophy of science: C. West Churchman,
'Elements of Logic and Formal Science', 1940; C. West
Churchman, 'Theory of Experimental Inference', 1948; Morris
R. Cohen and Ernest Nagel, 'An Introduction to Logic and
Scientific Method', 1934; A. Tarski, 'Introduction to Logic
and to the Methodology of Deductive Sciences', 1941.
22 P. Feyerabend, Consolations for the Specialist, in I. Lakatos
and A. Musgrave (eds), 'Criticism and the Growth of Knowledge', Cambridge University Press, 1970, p. 198.
23 Free Press, Chicago, 1950. See also: K. Dixon, 'Sociological
Theory: Pretence and Possibility', Routledge & Kegan Paul,
London and Boston, 1973; Ch.G.A. Bryant, 'Sociology in
Action: A Critique of Selected Conceptions of the Social Role
of the Sociologist, George Allen & Unwin, London, 1976.
24 A criticism of the use of Kuhn's ideas in the discussion of
sociology is to be found in J.D. Urry, Thomas S. Kuhn as
Sociologist of Knowledge, 'British Journal of Sociology',
vol. 24 (1973). When it comes to other social sciences, the
strongest impact of Kuhn's ideas can probably be observed
in political science. As has been shown by M.E. Kirn, post-
behaviouralism is largely a reflection of 'The Structure of
Scientific Revolutions':

Post-behaviorists ground their criticism of the behavioral
model of political science not in detailed analyses of
political inquiry, but rather in works which are critical
of some aspects of the logical empiricist and/or Popperian
concepts of science. In particular, post-behavioral critics
have relied upon T.S. Kuhn's Structure of Scientific
Revolutions . . . 'Behavioralism, Post-Behavioralism, and
the Philosophy of Science: Two Houses, One Plague, 'The
Review of Politics', vol. 39, January 1977, no. 1, p. 89).

It would be difficult to imagine a stronger similarity to that
which happens in sociology. The effects also are strikingly
convergent: 'Whether the point of view defended is
behavioralism or a particular version of post-behavioralism,
the effect is the same: the gap between practice and the
theory of that practice remains. Neither the behavioral

advocate nor the post-behavioral critic develops a concept of
political (italics, JDU) science (or contributes to the development of such a concept) *grounded in investigations of, and
reflections upon, either political inquiry or the theoretical
object of that inquiry, political activity.'* (author's italics)
(Ibid., pp. 87-8) Sociology leads in uncritically following
the philosophy of science, but is not the only discipline to do
so. When it comes to Kuhn's ideas, their impact can also be
observed in economics (see for instance Brian J. Loasby,
Hypothesis and Paradigm in the Theory of the Firm,
'Economic Journal', 81, December, 1971; Martin Bronfen-
brenner, The 'Structure of Revolutions' in Economic Thought,
'History of Political Economy', 3, Spring 1971; Leonard
Kunin and F. Stirton Weaver, On the Structure of Scientific
Revolutions in Economics, 'History of Political Economy', 3,
Fall 1971), and even in social anthropology (see Bob Scholte,
Epistemic Paradigms: Some Problems in Cross-Cultural
Research in Social Anthropological History and Theory,
'American Anthropologist', 68, 1966).
25 R.A.H. Robson, The Present State of Theory in Sociology,
in Imre Lakatos and Alan Musgrave (eds.), 'Problems in the
Philosophy of Science: Proceedings of the International
Colloquium in the Philosophy of Science', London, 1965, vol.
3, North-Holland Publishing Company, Amsterdam, 1968, pp.
369-70 (author's italics).
26 G. Casper Homans, A Life of Synthesis, in Irving L. Horowitz
(ed.), 'Sociological Self-Images', Sage, Beverly Hills, 1969,
p. 29.
27 The relationship between logical positivism and behaviourism
was, I think very accurately described by Carnap:

> The position we are advocating here coincides in its broad
> outlines with the philosophical movement known as 'behavior-
> ism' when, that is, its epistemological principles rather
> than its special methods are considered. We have not linked
> our exposition with a statement of behaviorism since our
> only concern is with epistemological foundations while
> behaviorism is above all else interested in a specific method
> of research and in specific concept formation. (R. Carnap,
> Psychology in Physical Language, in A.J. Ayer (ed.),
> 'Logical Positivism', Free Press, Chicago, 1956, p. 181.)

Carnap and other logical positivists in fact treated behavioural
psychology as one of the possible ways of putting the assumptions of logical positivism into effect in psychology. Carnap
himself was rather inclined to see the future of 'scientific'
psychology in connection with advances in neurophysiology.
  It was similar in the case of the relationship between
logical positivism and reductionism in Homans's version: 'the
general propositions of all the social sciences are propositions
of behavioral psychology' (cf. 'The Nature of Social Science',

Harcourt, Brace & World, New York, 1967, p. 108). That
strong version of reductionism is in agreement with the
epistemological foundations of logical positivism, but its
acceptance requires strong additional assumptions, which the
philosophers of science were usually unwilling to make.
28 G.C. Homans, The Sociological Relevance of Behaviorism,
in R.L. Burges, Don Bushell, Jr. (eds), 'Behavioral
Sociology: The Experimental Analysis of Social Process',
Columbia University Press, New York and London, 1969, p. 4.
29 'The so-called laws of supply and demand are certainly not
general. The demand for perfume, for instance, does not
obey the law: the higher the price of a perfume, the greater
the demand for it, at least up to a point' ('The Nature of
Social Science', op. cit., p. 20). Is the issue just that the
law of demand and supply, in the version adopted by Homans,
is false? Or that there are exceptions to that law? Not only
that. Here is the same example in another place:

> And most propositions about the aggregates, propositions
> that make no direct reference to the behavior of individuals,
> have no very wide generality; they hold good only within
> limited conditions. Perhaps some of the laws of economics,
> such as the so-called laws of supply and demand, come
> closest to being general. But they explicitly apply only
> within certain institutional conditions - markets and prices
> - and only within limits even there. 'The higher the price
> of a commodity, the less a consumer will buy' may not apply
> when high price gives the commodity great value as a status
> symbol (The Sociological Relevance of Behaviorism, op. cit.,
> p. 16).

Does the introduction of these additional limitations settle
the issue of the generality of the laws of supply and demand
in the negative sense? Homans obviously makes the fairly
common mistake of confusing general statements with uncon-
ditional ones. As a result, he is inclined to treat as general
statements in the social sciences only those which throw
light on the behaviour of *every* person, which may be useful
for the claim that 'the only general statements in the social
sciences are those of behavioral psychology' (hence arguments
about the merely alleged generality of the economic laws), but
has consequences which Homans himself would have to consider embarrassing, namely denying the status of generality
to those psychological laws (whether behavioural or not)
which state certain regularities under certain conditions, e.g.
conditions of an abundance of food. Be that as it may, Homans
uses the term 'general law' at variance with the meaning that
term has in the works on explanation he quotes. This is why
the arguments in favor of Hempel's conception of explanation
are not necessarily in favor of Homans's conception of explanation.

30 'The Nature of Social Science', op. cit., pp. 63-4. And here
   is a formulation which is even stronger in some respects:

> Let those who contend that behavioral psychology cannot
> explain the emergent characteristics of social behaviour
> [the opinion that comes close to what is really claimed] come
> forward and state the alternative general propositions that
> they think are necessary for this explanation. Let them
> show us their *deductive systems* with their *general pro-*
> *positions* in place. The fact is that they do not do so
> (The Sociological Relevance of Behaviorism, op. cit., p.
> 14. Author's italics).

31 'The Nature of Social Science', op. cit., p. 14. Note in this
   connection that for some reasons Boyle's law is the most
   popular example of a law of science in the eyes of the
   methodologists of the social sciences.
32 In fact his opinion of Marx has been quite good, at least
   since the time Marcuse won his popularity:

> I have tried to reduce the incidence of gulf in sociology,
> but the species responds to the antibiotic by breeding
> more virulent strains. Did we kill Marx to get Marcuse?
> What folly if we did! At times I suspect that both sense
> and nonsense have increased in sociology at the expense
> of some middle ground (A Life of Synthesis, op. cit.,
> p. 30).

33 As positive exceptions Homans classes the now rather for-
   gotten compilation of 'scientific discoveries' taken out of
   their original context, included in B. Berelson and G.A.
   Steiner, 'Human Behavior: An Inventory of Scientific Findings', Harcourt, Brace & World, New York, 1964.
34 'The Nature of Social Sciences', op. cit., p. 27, author's
   italics.
35 The Sociological Relevance of Behaviorism, op. cit., pp. 4-5,
   author's italics.
36 Cf. also my Methodological Dogma of Naturalism, op. cit.
37 First published in 1954 by Almqvist & Wiksells in Stockholm
   and by Tressler Press in New York.
38 Ibid., p. 73, author's italics.
39 Ibid., pp. 16-17.
40 Ibid., pp. 18-19. (In the version available to me propositions
   11 and 14 are identical.)
41 Ibid., p. 19.
42 N.C. Mullins, 'Theory and Theory Groups in Contemporary
   American Sociology', Harper & Row, New York, 1973, p. 218.
43 Cf. R.K. Merton, 'Social Theory and Social Structure',
   Free Press, Chicago, 1949; see also 'On Theoretical Sociology'
   by the same author, Free Press, New York, 1967.
44 J.P. Gibbs, 'Sociological Theory Construction', The Dryden

Press, Hinsdale, Ill., 1972, p. vi. See also: J. Hage, 'Techniques and Problems of Theory Construction in Sociology', John Wiley, New York, 1972; Ch. Winton, 'Theory and Measurement in Sociology', John Wiley, New York, 1974.

45 H.M. Blalock, Jr., 'Theory Construction: From Verbal to Mathematical Formulations', Prentice-Hall, Englewood Cliffs, New Jersey, 1969, p. 3. See also H.M. Blalock, Jr., and A.B. Blalock (eds), 'Methodology in Social Research', McGraw Hill, New York, 1968.

46 This, of course, applies not only to sociology. One of the latest handbooks of methodology of research for students of political science, defines theory as 'a set of empirical generalizations (or hypotheses or laws) that are connected deductively', and states further that

> The definition of 'theory' implies that 'deducibility' is a defining characteristic of theory. For a set of statements to be labelled appropriately as a 'theory', it must satisfy the requirement of being arranged deductively. Theories, in other words, must take the form of an axiomatic system.

The authors of the handbook further explain the structure of an axiomatic system by referring to school-level geometry and give an example of an axiomatic system in the field of the social sciences, namely - of course! - 'Durkheim's theory of the division of labor' as interpreted by Zetterberg. The chapter on theory is concluded by the learning aids, the first three of which are given below:

> 1 A theory is a set of empirical . . . that are connected deductively.
> 2 Theories take the form of . . . systems.
> 3 Thus, theories are sets of empirical generalizations that are connected.

The learning aids also include an exercise concerned with constructing an axiomatic system out of four given axioms. Cf. D. McGaw and G. Watson, 'Political and Social Inquiry', John Wiley, New York, London, Sydney and Toronto, 1976, pp. 170-3.

47 The book by Berelson and Steiner (op. cit.) is probably the worst possible example. And here are some other titles: K.P. Schwirian and J.W. Prehn, An Axiomatic Theory of Urbanization, 'American Sociological Review', 27 (December 1962); L.C. Gould and C. Schrag, Theory Construction and Prediction of Juvenile Delinquency, 'Proceedings of the Social Statistics Section of the American Statistical Association', 1962; J.P. Gibbs and W.T. Martin, Urbanization, Technology, and the Division of Labor: International Patterns, 'American Sociological Review', 27 (October 1962); W.R. Catton, Jr., The Functions and Dysfunctions of Ethnocentrism

A Theory, 'Social Problems', 8 (Winter 1961); J.P. Gibbs
and W.P. Martin, 'Status Integration and Suicide: A Sociological Study', University of Oregon Books, 1964. A
criticism of the first item, and indirectly the whole production
of this kind, made from empiricist positions is to be found
in O.D. Duncan Axioms or Correlations?, 'American Sociological Review', 28 (June 1963).

48 This is excellently presented, with reference to the early
period of the evolution of logical positivism, in the now
classic study by Janina Kotarbińska, Ewolucja Koła Wiedeńs-
kiego (The Evolution of the Vienna Circle), Myśl Współczesna,
2, 1947.

49 Cf. C. G. Hempel, The Empiricist Criterion of Meaning, in
A.J. Ayer (ed.), 'Logical Positivism', Free Press, Chicago,
1959; C.G. Hempel, 'Aspects of Scientific Explanation', Free
Press, New York, 1965. The literature on the subject of
theoretical terms is, of course, very comprehensive. See
R. Tuomela, 'Theoretical Concepts', Springer-Verlag, Wien-
New York, 1973, which is a comprehensive monograph of the
subject, included in The Library of Exact Philosophy.

50 Cf. G.A. Lundberg, Operational Definitions in the Social
Sciences, 'American Journal of Sociology', vol. xlvii, 1942,
pp. 131ff. The protest was formulated, of course, in connection with reflections intended to add precision to the terms
being used. Note also that Dodd let himself be outmanoeuvred
by his opponents and in the next issue of 'American Journal
of Sociology' defined the operational definition operationally
(cf. S. Dodd, Operational Definitions Operationally Defined,
'American Journal of Sociology', Vol. xlviii, 1943).

51 Among the leading logical positivists, Carnap probably came
closest to identifying operationism with the corresponding
elements of the logical-positivist doctrine. He claimed that
'the requirements of testability and of operationism as
represented by various authors are closely related to each
other, differing only in minor details and in emphasis.'
Carnap also thought that

> the principle of operationism, which was first proposed in
> physics by Bridgman and then applied also in other fields
> of science, including psychology, had on the whole a
> healthy effect on the procedure of concept formation used
> by scientists. The principle has contributed to the clarifi-
> cation of many concepts and has helped to eliminate unclear
> or even unscientific concepts. (The Methodological Character
> of Theoretical Concepts, 'Minnesota Studies in the
> Philosophy of Science', I, University of Minnesota, Minnea-
> polis, 1965, p. 65).

52 One of the best analyses of operationism and its consequences
is to be found in M. Przełęcki's paper O tzw. definicjach
operacyjnych (On so-called Operational Definitions) 'Studia

Logica', III, 1955, and Operacjonizm (Operationism)
'Archiwum Historii Filozofii i Myśli Społecznej', vol. V, 1959.
When it comes to the discussion of operationism in the social
sciences, the following items deserve attention: R. Bain, On
Attitude and Attitude Research, 'American Journal of Soci-
ology', vol. 33, 1928; R. Bain, Die Behavioristische
Einstellung in der Soziologie, 'Sociologus', IX, 1933; Read
Bain, Measurement in Sociology, 'American Journal of
Sociology', vol. 40, 1935; S.S. Stevens, The Operational
Basis of Psychology, 'American Journal of Psychology', 47,
1934; G. Lundberg, 'Foundations of Sociology', Macmillan,
New York, 1939; G. Lundberg, The Thoughtways of Con-
temporary Sociology, 'American Sociological Review', I, 1936;
G. Lundberg, The Measurement of Socioeconomic Status,
'American Sociological Review', V, 1940; George Lundberg,
'Social Research', Longmans, Green, New York, 1942; G.
Lundberg, Operational Definitions in the Social Sciences,
'American Journal of Sociology', XLVII, 1942; H. Blumer,
Rejoinder to Lundberg's 'Operational Definitions in the Social
Sciences', 'American Journal of Sociology', XLVII, 1942; H.
Blumer, The Problem of the Concept in Social Psychology,
'American Journal of Sociology', XLV, 1940; A. Allpert,
Operational Definitions in Sociology, 'American Sociological
Review', III, 1938; S.C. Dodd, A System of Operationally
Defined Concepts for Sociology, 'American Sociological Review',
IV, 1939; S.C. Dodd, A Tension Theory of Societal Action,
'American Sociological Review', iv, 1939; S.C. Dodd,
Operational Definitions Operationally Defined, 'American
Journal of Sociology', XLVIII, 1943; E. Shanas, Comment to
Stuart S. Dodd's, 'Operational Definitions Operationally
Defined', 'American Journal of Sociology', XLVIII, 1943; F.S
Chapin, Definition of Definition of Concepts, 'Social Forces',
XVIII, 1939; Clifford Kirkpatric, A Methodological Analysis
of Feminism in Relation to Marital Adjustment, 'American
Sociological Review', IV, 1939; G. Allport, The Psychologist'
Frame of Reference, 'Psychological Bulletin', XXXVII, 1940;
the entire issue of the 'Psychological Review', LII, 5, 1945;
F. Adler, Operational Definitions in Sociology, 'American
Journal of Sociology', LII, 1947; E.G. Boring, 'History of
Experimental Psychology', 2nd ed, 1950; Hornell Hart and
Associates, Toward an Operational Definition of the Term
'Operation', 'American Sociological Review', XVIII, 1953.
53 P.W. Bridgman, Remarks on the Present State of Operationalism, 'Scientific Monthly', 79 (October 1954), p. 224. Quoted
from Gideon Sjeberg, Operationalism and Social Research, in
L. Gross (ed.), 'Symposium on Social Theory', Row, Peter-
son, Evanston, Ill., 1959, p. 604.
54 H. Zetterberg, 'On Theory and Verification in Sociology',
op. cit., p. 34.
55 Ibid.
56 Ibid., p. 30.

7 G.A. Lundberg, 'Foundations of Sociology', Macmillan, New York, 1939, p. 59.

8 H. Zetterberg, op. cit., p. 30.

9 Ibid., p. 36.

0 Cf. W.J. Good and P.K. Hatt, 'Methods in Social Research', McGraw-Hill, New York, Toronto, London, 1952.

1 Ibid., p. 237.

2 'Założenia socjologii humanistycznej' (Assumptions of Humanistic Sociology), PWN, Warsaw, 1971.

3 H. Zetterberg, op. cit., pp. 34-5.

4 C. Selltiz et al., 'Research Methods in Social Relations', revised one-volume edition, Holt, Rinehart & Winston, New York, 1964, p. 41.

5 M. Marody, 'Sens teoretyczny a sens empiryczny pojęcia postawy' (The Theoretical Versus the Empirical Meaning or the Concept of Attitude), PWN, Warsaw, 1976, p. 9. Author's italics.

6 Kaplan's paper Definitions and Specification of Meaning, 'The Journal of Philosophy', vol. xliii, no. 11, May 1946, which in some respects is ahead of The Methodological Character of Theoretical Concepts, by Carnap (1959), is reprinted in the well-known book 'The Language of Social Research', edited by P.F. Lazarsfeld and M. Rosenberg (1955); Kaplan's book 'The Conduct of Inquiry' (1964), liberalizes the requirements addressed to the language of science more than is possible in logical positivism; it functions as a handbook of the philosophy of science for sociologists.

7 R.G. Dumont and W.J. Wilson, Aspects of Concept Formation, Explication and Theory Construction in Sociology, 'American Sociological Review', vol. 32, no. 6, December 1967.

8 Ibid., p. 985.

9 Ibid., p. 986. (Author's italics.)

0 Ibid., p. 987.

1 Ibid.

2 C. Hempel, The Function of General Laws in History, in Patric Gardiner (ed.), 'Theories of History', Free Press, Chicago, 1959.

3 Dumont and Wilson, op. cit., p. 988.

4 Ibid., p. 989.

5 Ibid., p. 994.

6 A. Kaplan, 'Conduct of Inquiry', op. cit., pp. 56-7.

7 The text by Dumont and Wilson was included in the anthology edited by D.P. Forcese and S. Richter and bearing the telling title 'Stages of Social Research: Contemporary Perspectives' (Prentice-Hall, Englewood Cliffs, 1970). The same anthology also included a paper by Allan Mazur, The Littlest Science, in which its author diagnoses sociology as revealed in the title and seriously recommends, as the only remedy, that sociologists accumulate as much new data as possible, because the history of science shows (here follows a display of the knowledge of the history of physics over the last four hundred

years, so typical of sociologists) that great theoretical discoveries originate from new observations. *'Physics has had a*
continual flow of new data - sometimes a flood of it. *Sociolog:*
has not, and that is at least one reason why it is such a pun
science p. 7, (italics in the original). And already in the
late 1950s Dennis Wrong thought that 'no one believes any
longer that "the facts speak for themselves"'. (The Failure
of American Sociology, 'Commentary', vol. XXVIII, no. 5,
November 1959.)

78 Ossowska claimed that, judging from the way in which the
term 'cognitive dissonance' is used, it denotes all that it
should denote to leave the main theses of the theory of cogni
tive dissonance unimpaired. Cf. M. Ossowska, Uwagi o
pojęciu dysonansu poznawczego u L. Festingera (Comments
on the Concept of Cognitive Dissonance in L. Festinger),
'Studia Socjologiczne', no. 1 (14), 1962.

79 A. Malewski, W sprawie dyskusji nad teorią dysonansu
poznawczego (On the Discussion of the Theory of Cognitive
Dissonance), 'Studia Socjologiczne', no. 2 (5), 1962. The
quotations are from pp. 236-7.

80 A. Malewski, Empiryczny sens teorii materializmu historyczn
(The Empirical Sense of the Theory of Historical Materialism)
'Studia Filozoficzne', no. 2, 1957.

81 Ibid., pp. 71-2.

82 Ibid., p. 81.

83 Cf. D. Nachmias and Ch. Nachmias, 'Research Methods in th
Social Sciences', St Martin's Press, New York, 1976. The
opening chapters of that handbook (Chapter 1: The Scientifi
Methods; Chapter 2: Basic Elements of Research; Chapter 3:
The Research Design) are marked by the lack of certainty
typical of texts written by non-professionals. See also S.
Olson, 'Ideas and Data: The Process and Practice of Social
Research', Dorsey Press, Homewood, Ill., 1976. To the
reader in search of a clearly incompetent and naive methodo-
logical text I recommend the introduction to 'Sociological
Theory: An Introduction', W.L. Wallace (ed.), Aldine,
Chicago, 1969.

6 EXAMPLE I: PAUL LAZARSFELD: FROM CONCEPTS TO
INDICATORS

1 Cf. Jerzy Szacki, 'History of Sociological Thought', Green-
wood Press, West Port, Connecticut, 1979.

2 Martin Jay, 'The Dialectical Imagination: A History of the
Frankfurt School and the Institute of Social Research, 1923-
1950', Little, Brown, Boston, Toronto, 1973.

3 See on this subject Włodzimierz Wesołowski, 'Teoria, badania
praktyka: z problematyki struktury klasowej' (Theory,
Research, Practice: Issues in Class Structure), Książka i
Wiedza, Warszawa, 1975.

4 Paul F. Lazarsfeld and Morris Rosenberg (eds), 'The
  Language of Social Research: A Reader in the Methodology
  of Social Research', The Free Press of Glencoe, 3rd edn,
  1962, p. 2. Formally both editors are the authors of the (not
  undersigned) introduction, but in fact we can treat Lazarsfeld as the author of the general introduction and the introduction to Part I (i.e., of those texts in which we are
  primarily interested here), since those texts coincide with
  other writings by Lazarsfeld. For instance, see following
  items by that author: Methodological Problems in Empirical
  Social Research, 'Transactions of the Fourth World Congress
  of Sociology', vol. II, International Sociological Association,
  1959, Evidence and Inference in Social Research, 'Daedalus',
  vol. 87, no. 4, 1958; The Place of Empirical Social Research
  in the Map of Contemporary Sociology, in John McKinney
  and Edward A. Tiryakian (eds), 'Theoretical Sociology:
  Perspectives and Developments', Appleton-Century-Crofts,
  New York, 1970.
5 Ibid. (author's italics).
6 Paul F. Lazarsfeld, Evidence and Inference in Social Research,
  op. cit., p. 101.
7 Ibid.
8 Ibid., p. 102. (Author's italics.)
9 Ibid., p. 103.
10 Ibid., p. 104.
11 Ibid., pp. 104-5.
12 Cf. Halina Mortimer, Logiczne podstawy 'Analizy ukrytej
  struktury' (The Logical Foundations of 'Latent Structure
  Analysis'), 'Studia Filozoficzne', 2, 1968.
13 Stefan Nowak, Czy analiza struktury ukrytej jest metodą
  pomiaru? (Is Latent Structure Analysis a Method of Measurement?) in Stefan Nowak (ed.), 'Metody badań socjologicznych'
  ('Methods of Sociological Research') PWN, Warszawa, 1965,
  p. 327.
14 Cf. 'The Level of Living Index', New Version, UNRISD,
  Geneva, 1968.
15 Cf. Leszek Zienkowski, Czy potrafimy mierzyć 'poziom życia'?
  (Can we Measure the 'Level of Living?') 'Wiadomości Staty-
  styczne', no. 11, 1973. See also 'Problemy mierników poziomu
  życia ludności' (Problems of Measures of the Level of Living),
  GUS, Warszawa, 1974.
16 'The Language of Social Research', op. cit., p. 4.
17 Stefan Nowak, Studia z metodologii nauk społecznych (Studies
  in the Methodology of the Social Sciences), PWN, Warszawa,
  1965, pp. 245-6.

7 EXAMPLE II: THE METHODOLOGY OF COMPARATIVE STUDIES

1 James S. Coleman, The Methods of Sociology, in R. Bierstedt
  (ed.), 'A Design for Sociology: Scope, Objectives, and

Methods', American Academy of Political and Social Science, Monograph 9, Philadelphia, April, 1969. Quoted from the reprint in: D.P. Forcese and S. Richter (eds), 'Stages of Social Research: Contemporary Perspectives', Prentice-Hall, Englewood Cliffs, New Jersey, 1970, p. 408.

2 See e.g. Stein Rokkan et al. (ed.), 'Comparative Analysis, Mouton, The Hague, Paris, 1969; Adam Przeworski and Henry Teune, 'The Logic of Comparative Social Inquiry', Wiley, New York, 1970; Ivan Vallier (ed.), 'Comparative Methods in Sociology', University of California Press, Berkeley, Los Angeles, London, 1971; Alexander Szalai and Ricardo Petrella (eds), 'Cross-National Comparative Survey Research', International Social Science Council, Pergamon Press, Oxford, New York, Toronto, Sydney, Paris, Frankfurt, 1977.

3 Sjoerd Groenman, foreword to 'Values and the Active Community', International Studies of Values in Politics, Free Press, New York, 1971, p. xv.

4 Robert M. Marsh, Comparative Sociology 1950-1963, 'Current Sociology', XIV (1966), no. 2, p. 5.

5 Marion J. Levy, Jr., Scientific Analysis is a Subset of Comparative Analysis, in John C. McKinney and Edward A. Tiryakian (eds), 'Theoretical Sociology', Appleton-Century-Crofts, New York, 1970.

6 Morris Zelditch, Jr., Intelligible Comparisons, in Ivan Vallier (ed.), 'Comparative Methods in Sociology', ed. cit., p. 271.

7 Ibid., p. 267.

8 Ibid., p. 268.

9 Adam Przeworski and Henry Teune, 'The Logic of Comparative Social Inquiry', op. cit.

10 Ibid., p. xi.

11 Ibid., p. 17.

12 Carl. G. Hempel and Paul Oppenheim, The Logic of Explanation in Herbert Feigl and May Brodbeck (eds), 'Readings in the Philosophy of Science', Appleton-Century-Crofts, New York, 1953, p. 338.

13 Cf. Andrzej Malewski and Jerzy Topolski, 'Studia z metodologii historii' ('Studies in the Methodology of History'), PWN, Warszawa, 1960.

14 Cf. Stanisław Ossowski, 'Dzieła', (Collected Works), vol. IV, PWN, Warszawa, 1967.

15 'The Logic of Comparative Social Inquiry', op. cit., p. 4 (author's italics).

16 Cf. 'Values and the Active Community'.

17 Robert M. Marsh, Comparative Sociology, 1950-1963, op. cit., p. 6.

18 'The Logic of Comparative Social Inquiry', op. cit., p. 10.

19 Ibid., p. 8 (author's italics).

20 Ibid., p. 24.

21 As far as I know, this opinion was first formulated by Stefan

Nowak in his paper Prawa ogólne i generalizacje historyczne
w naukach społecznych (General Laws and Historical Generalizations in the Social Sciences), published inter alia in: Stefan
Nowak, 'Studia z metodologii nauk społecznych', (Studies
in the Methodology of the Social Sciences), PWN, Warszawa,
1965. See also The Strategy of Cross-National Survey
Research for the Development of Social Theory by the same
author, published in Alexander Szalai and Ricardo Petrella
(eds), 'Cross-National Comparative Survey Research', op. cit.
22 'The Logic of Comparative Social Inquiry', op. cit., p. 29.
23 Ibid., p. 23.
24 Ibid., p. 17.
25 Charles Taylor, Interpretation and the Science of Man, 'The
Review of Methaphysics', no. 1, p. 34.

POSTSCRIPT

1 Cf. Raoul Naroll and Ronald Cohen (eds), 'A Handbook of
Method in Cultural Anthropology', Columbia University
Press, New York and London, 1973.
2 This applies, incidentally, not only to the humanities and the
social sciences, for it also covers some other 'less advanced'
sciences. By the way of example we can mention David Harvey,
'Explanation in Geography', Edward Arnold, London, 1969.
In his review of that book Robert Sack wrote: 'what is most
striking about the book is that it has very little to do with
geography.' ('Historical Methods Newsletter', vol. 6, no. 2,
March 1963, p. 68.) Is that really so striking?
3 J. Rogers Hollingsworth, Some Problems in Theory Construction for Historical Analysis, 'Historical Methods Newsletter',
vol. 7, no. 3, June 1974, p. 226.
4 Ibid.
5 Ibid., p. 227.
6 Ibid., p. 228.
7 Ibid., p. 234.
8 Erik Allardt, A Comment on Hollingsworth, 'Historical
Methods Newsletter', vol. 7, no. 3, June 1974, p. 245.

Index

Routledge Social Science Series

Routledge & Kegan Paul   London, Henley and Boston

39 Store Street,
London WC1E 7DD
Broadway House,
Newtown Road,
Henley-on-Thames,
Oxon RG9 1EN
9 Park Street,
Boston, Mass. 02108

Contents

*Authors wishing to submit manuscripts for any series
in this catalogue should send them to the Social Science Editor,
Routledge & Kegan Paul Ltd, 39 Store Street,
London WC1E 7DD.*
● *Books so marked are available in paperback.*
○ *Books so marked are available in paperback only.*
*All books are in metric Demy 8vo format (216 × 138mm approx.)
unless otherwise stated.*

International Library of Sociology
General Editor   John Rex

GENERAL SOCIOLOGY

Barnsley, J. H. The Social Reality of Ethics. *464 pp.*
Brown, Robert. Explanation in Social Science. *208 pp.*
● Rules and Laws in Sociology. *192 pp.*
Bruford, W. H. Chekhov and His Russia. *A Sociological Study. 244 pp.*
Burton, F. and **Carlen, P.** Official Discourse. *On Discourse Analysis, Government Publications, Ideology. About 140 pp.*
Cain, Maureen E. Society and the Policeman's Role. *326 pp.*
● **Fletcher, Colin.** Beneath the Surface. *An Account of Three Styles of Sociological Research. 221 pp.*
Gibson, Quentin. The Logic of Social Enquiry. *240 pp.*
Glassner, B. Essential Interactionism. *208 pp.*
Glucksmann, M. Structuralist Analysis in Contemporary Social Thought. *212 pp.*
Gurvitch, Georges. Sociology of Law. *Foreword by Roscoe Pound. 264 pp.*
Hinkle, R. Founding Theory of American Sociology 1881–1913. *About 350 pp.*
Homans, George C. Sentiments and Activities. *336 pp.*
Johnson, Harry M. Sociology: *A Systematic Introduction. Foreword by Robert K. Merton. 710 pp.*
● **Keat, Russell** and **Urry, John.** Social Theory as Science. *278 pp.*
Mannheim, Karl. Essays on Sociology and Social Psychology. *Edited by Paul Keckskemeti. With Editorial Note by Adolph Lowe. 344 pp.*
Martindale, Don. The Nature and Types of Sociological Theory. *292 pp.*
● **Maus, Heinz.** A Short History of Sociology. *234 pp.*
Myrdal, Gunnar. Value in Social Theory: *A Collection of Essays on Methodology. Edited by Paul Streeten. 332 pp.*
Ogburn, William F. and **Nimkoff, Meyer F.** A Handbook of Sociology. *Preface by Karl Mannheim. 656 pp. 46 figures. 35 tables.*
Parsons, Talcott and **Smelser, Neil J.** Economy and Society: *A Study in the Integration of Economic and Social Theory. 362 pp.*
Payne, G., Dingwall, R., Payne, J. and **Carter, M.** Sociology and Social Research. *About 250 pp.*
Podgórecki, A. Practical Social Sciences. *About 200 pp.*
Podgórecki, A. and **Łos, M.** Multidimensional Sociology. *268 pp.*
Raffel, S. Matters of Fact. *A Sociological Inquiry. 152 pp.*
● **Rex, John.** Key Problems of Sociological Theory. *220 pp.*
Sociology and the Demystification of the Modern World. *282 pp.*
● **Rex, John.** (Ed.) Approaches to Sociology. *Contributions by Peter Abell, Frank Bechhofer, Basil Bernstein, Ronald Fletcher, David Frisby, Miriam Glucksmann, Peter Lassman, Herminio Martins, John Rex, Roland Robertson, John Westergaard and Jock Young. 302 pp.*
Rigby, A. Alternative Realities. *352 pp.*
Roche, M. Phenomenology, Language and the Social Sciences. *374 pp.*
Sahay, A. Sociological Analysis. *220 pp.*
Strasser, Hermann. The Normative Structure of Sociology. *Conservative and Emancipatory Themes in Social Thought. About 340 pp.*
Strong, P. Ceremonial Order of the Clinic. *267 pp.*
Urry, John. Reference Groups and the Theory of Revolution. *244 pp.*
Weinberg, E. Development of Sociology in the Soviet Union. *173 pp.*

FOREIGN CLASSICS OF SOCIOLOGY

● **Gerth, H. H.** and **Mills, C. Wright.** From Max Weber: *Essays in Sociology. 502 pp.*

3

● **Tönnies, Ferdinand.** Community and Association *(Gemeinschaft und Gesellschaft).|Translated and Supplemented by Charles P. Loomis. Foreword by Pitirim A. Sorokin. 334 pp.*

SOCIAL STRUCTURE

Andreski, Stanislav. Military Organization and Society. *Foreword by Professor A. R. Radcliffe-Brown. 226 pp. 1 folder.*

Broom, L., Lancaster Jones, F., McDonnell, P. and **Williams, T.** The Inheritance of Inequality. *About 180 pp.*

Carlton, Eric. Ideology and Social Order. *Foreword by Professor Philip Abrahams. About 320 pp.*

Clegg, S. and **Dunkerley, D.** Organization, Class and Control. *614 pp.*

Coontz, Sydney H. Population Theories and the Economic Interpretation. *202 pp.*

Coser, Lewis. The Functions of Social Conflict. *204 pp.*

Crook, I. and **D.** The First Years of the Yangyi Commune. *304 pp., illustrated.*

Dickie-Clark, H. F. Marginal Situation: *A Sociological Study of a Coloured Group. 240 pp. 11 tables.*

Giner, S. and **Archer, M. S.** (Eds) Contemporary Europe: *Social Structures and Cultural Patterns, 336 pp.*

● **Glaser, Barney** and **Strauss, Anselm L.** Status Passage: *A Formal Theory. 212 pp.*

Glass, D. V. (Ed.) Social Mobility in Britain. *Contributions by J. Berent, T. Bottomore, R. C. Chambers, J. Floud, D. V. Glass, J. R. Hall, H. T. Himmelweit, R. K. Kelsall, F. M. Martin, C. A. Moser, R. Mukherjee and W. Ziegel. 420 pp.*

Kelsall, R. K. Higher Civil Servants in Britain: *From 1870 to the Present Day. 268 pp. 31 tables.*

● **Lawton, Denis.** Social Class, Language and Education. *192 pp.*

McLeish, John. The Theory of Social Change: *Four Views Considered. 128 pp.*

● **Marsh, David C.** The Changing Social Structure of England and Wales, 1871–1961. *Revised edition. 288 pp.*

Menzies, Ken. Talcott Parsons and the Social Image of Man. *About 208 pp.*

● **Mouzelis, Nicos.** Organization and Bureaucracy. *An Analysis of Modern Theories. 240 pp.*

● **Ossowski, Stanislaw.** Class Structure in the Social Consciousness. *210 pp.*

● **Podgórecki, Adam.** Law and Society. *302 pp.*

Renner, Karl. Institutions of Private Law and Their Social Functions. *Edited, with an Introduction and Notes, by O. Kahn-Freud. Translated by Agnes Schwarzschild. 316 pp.*

Rex, J. and **Tomlinson, S.** Colonial Immigrants in a British City. *A Class Analysis. 368 pp.*

Smooha, S. Israel: Pluralism and Conflict. *472 pp.*

Wesolowski, W. Class, Strata and Power. *Trans. and with Introduction by G. Kolankiewicz. 160 pp.*

Zureik, E. Palestinians in Israel. *A Study in Internal Colonialism. 264 pp.*

SOCIOLOGY AND POLITICS

Acton, T. A. Gypsy Politics and Social Change. *316 pp.*

Burton, F. Politics of Legitimacy. *Struggles in a Belfast Community. 250 pp.*

Crook, I. and **D.** Revolution in a Chinese Village. *Ten Mile Inn. 216 pp., illustrated.*

Etzioni-Halevy, E. Political Manipulation and Administrative Power. *A Comparative Study. About 200 pp.*

Fielding, N. The National Front. *About 250 pp.*

● **Hechter, Michael.** Internal Colonialism. *The Celtic Fringe in British National Development, 1536–1966. 380 pp.*

Kornhauser, William. The Politics of Mass Society. *272 pp. 20 tables.*

Korpi, W. The Working Class in Welfare Capitalism. *Work, Unions and Politics in Sweden. 472 pp.*

Kroes, R. Soldiers and Students. *A Study of Right- and Left-wing Students. 174 pp.*

Martin, Roderick. Sociology of Power. *About 272 pp.*

Merquior, J. G. Rousseau and Weber. *A Study in the Theory of Legitimacy. About 288 pp.*

Myrdal, Gunnar. The Political Element in the Development of Economic Theory. *Translated from the German by Paul Streeten. 282 pp.*

Varma, B. N. The Sociology and Politics of Development. *A Theoretical Study. 236 pp.*

Wong, S.-L. Sociology and Socialism in Contemporary China. *160 pp.*

Wootton, Graham. Workers, Unions and the State. *188 pp.*

CRIMINOLOGY

Ancel, Marc. Social Defence: *A Modern Approach to Criminal Problems. Foreword by Leon Radzinowicz. 240 pp.*

Athens, L. Violent Criminal Acts and Actors. *104 pp.*

Cain, Maureen E. Society and the Policeman's Role. *326 pp.*

Cloward, Richard A. and **Ohlin, Lloyd E.** Delinquency and Opportunity: *A Theory of Delinquent Gangs. 248 pp.*

Downes, David M. The Delinquent Solution. *A Study in Subcultural Theory. 296 pp.*

Friedlander, Kate. The Psycho-Analytical Approach to Juvenile Delinquency: *Theory, Case Studies, Treatment. 320 pp.*

Gleuck, Sheldon and **Eleanor.** Family Environment and Delinquency. *With the statistical assistance of Rose W. Kneznek. 340 pp.*

Lopez-Rey, Manuel. Crime. *An Analytical Appraisal. 288 pp.*

Mannheim, Hermann. Comparative Criminology: *A Text Book. Two volumes. 442 pp. and 380 pp.*

Morris, Terence. The Criminal Area: *A Study in Social Ecology. Foreword by Hermann Mannheim. 232 pp. 25 tables. 4 maps.*

Rock, Paul. Making People Pay. *338 pp.*

● **Taylor, Ian, Walton, Paul** and **Young, Jock.** The New Criminology. *For a Social Theory of Deviance. 325 pp.*

● **Taylor, Ian, Walton, Paul** and **Young, Jock.** (Eds) Critical Criminology. *268 pp.*

SOCIAL PSYCHOLOGY

Bagley, Christopher. The Social Psychology of the Epileptic Child. *320 pp.*

Brittan, Arthur. Meanings and Situations. *224 pp.*

Carroll, J. Break-Out from the Crystal Palace. *200 pp.*

● **Fleming, C. M.** Adolescence: Its Social Psychology. *With an Introduction to recent findings from the fields of Anthropology, Physiology, Medicine, Psychometrics and Sociometry. 288 pp.*

● The Social Psychology of Education: *An Introduction and Guide to Its Study. 136 pp.*

Linton, Ralph. The Cultural Background of Personality. *132 pp.*

● **Mayo, Elton.** The Social Problems of an Industrial Civilization. *With an Appendix on the Political Problem. 180 pp.*

Ottaway, A. K. C. Learning Through Group Experience. *176 pp.*

Plummer, Ken. Sexual Stigma. *An Interactionist Account. 254 pp.*

● **Rose, Arnold M.** (Ed.) Human Behaviour and Social Processes: *an Interactionist Approach. Contributions by Arnold M. Rose, Ralph H. Turner, Anselm Strauss, Everett C. Hughes, E. Franklin Frazier, Howard S. Becker et al. 696 pp.*

Smelser, Neil J. Theory of Collective Behaviour. *448 pp.*

Stephenson, Geoffrey M. The Development of Conscience. *128 pp.*

Young, Kimball. Handbook of Social Psychology. *658 pp. 16 figures. 10 tables.*

SOCIOLOGY OF THE FAMILY

Bell, Colin R. Middle Class Families: *Social and Geographical Mobility. 224 pp.*
Burton, Lindy. Vulnerable Children. *272 pp.*
Gavron, Hannah. The Captive Wife: *Conflicts of Household Mothers. 190 pp.*
George, Victor and **Wilding, Paul.** Motherless Families. *248 pp.*
Klein, Josephine. Samples from English Cultures.
   1. Three Preliminary Studies and Aspects of Adult Life in England. *447 pp.*
   2. Child-Rearing Practices and Index. *247 pp.*
Klein, Viola. The Feminine Character. *History of an Ideology. 244 pp.*
McWhinnie, Alexina M. Adopted Children. *How They Grow Up. 304 pp.*
● **Morgan, D. H. J.** Social Theory and the Family. *About 320 pp.*
● **Myrdal, Alva** and **Klein, Viola.** Women's Two Roles: *Home and Work. 238 pp.*
   27 tables.
Parsons, Talcott and **Bales, Robert F.** Family: Socialization and Interaction Process.
   In collaboration with James Olds, Morris Zelditch and Philip E. Slater. 456 pp.
   50 figures and tables.

SOCIAL SERVICES

Bastide, Roger. The Sociology of Mental Disorder. *Translated from the French by Jean McNeil. 260 pp.*
Carlebach, Julius. Caring For Children in Trouble. *266 pp.*
George, Victor. Foster Care. *Theory and Practice. 234 pp.*
   Social Security: *Beveridge and After. 258 pp.*
George, V. and **Wilding, P.** Motherless Families. *248 pp.*
● **Goetschius, George W.** Working with Community Groups. *256 pp.*
Goetschius, George W. and **Tash, Joan.** Working with Unattached Youth. *416 pp.*
Heywood, Jean S. Children in Care. *The Development of the Service for the Deprived Child. Third revised edition. 284 pp.*
King, Roy D., Ranes, Norma V. and **Tizard, Jack.** Patterns of Residential Care. *356 pp.*
Leigh, John. Young People and Leisure. *256 pp.*
● **Mays, John.** (Ed.) Penelope Hall's Social Services of England and Wales. *368 pp.*
Morris, Mary. Voluntary Work and the Welfare State. *300 pp.*
Nokes, P. L. The Professional Task in Welfare Practice. *152 pp.*
Timms, Noel. Psychiatric Social Work in Great Britain (1939–1962). *280 pp.*
●   Social Casework: *Principles and Practice. 256 pp.*

SOCIOLOGY OF EDUCATION

Banks, Olive. Parity and Prestige in English Secondary Education: a Study in Educational Sociology. *272 pp.*
● **Blyth, W. A. L.** English Primary Education. *A Sociological Description.*
   2. Background. *168 pp.*
Collier, K. G. The Social Purposes of Education: *Personal and Social Values in Education. 268 pp.*
Evans, K. M. Sociometry and Education. *158 pp.*
● **Ford, Julienne.** Social Class and the Comprehensive School. *192 pp.*
Foster, P. J. Education and Social Change in Ghana. *336 pp. 3 maps.*
Fraser, W. R. Education and Society in Modern France. *150 pp.*
Grace, Gerald R. Role Conflict and the Teacher. *150 pp.*
Hans, Nicholas. New Trends in Education in the Eighteenth Century. *278 pp.*
   19 tables.
●   Comparative Education: *A Study of Educational Factors and Traditions. 360 pp.*
● **Hargreaves, David.** Interpersonal Relations and Education. *432 pp.*
●   Social Relations in a Secondary School. *240 pp.*
   School Organization and Pupil Involvement. *A Study of Secondary Schools.*

● **Mannheim, Karl** and **Stewart, W. A. C.** An Introduction to the Sociology of
  Education. *206 pp.*
● **Musgrove, F.** Youth and the Social Order. *176 pp.*
● **Ottaway, A. K. C.** Education and Society: An Introduction to the Sociology of
  Education. *With an Introduction by W. O. Lester Smith. 212 pp.*
  Peers, Robert. Adult Education: *A Comparative Study. Revised edition. 398 pp.*
  Stratta, Erica. The Education of Borstal Boys. *A Study of their Educational
  Experiences prior to, and during, Borstal Training. 256 pp.*
● **Taylor, P. H., Reid, W. A.** and **Holley, B. J.** The English Sixth Form. *A Case Study in
  Curriculum Research. 198 pp.*

SOCIOLOGY OF CULTURE

Eppel, E. M. and **M.** Adolescents and Morality: *A Study of some Moral Values and
  Dilemmas of Working Adolescents in the Context of a changing Climate of
  Opinion. Foreword by W. J. H. Sprott. 268 pp. 39 tables.*
● **Fromm, Erich.** The Fear of Freedom. *286 pp.*
●   The Sane Society. *400 pp.*
  Johnson, L. The Cultural Critics. *From Matthew Arnold to Raymond Williams.
  233 pp.*
  Mannheim, Karl. Essays on the Sociology of Culture. *Edited by Ernst Mannheim in
  co-operation with Paul Kecskemeti. Editorial Note by Adolph Lowe. 280 pp.*
  Merquior, J. G. The Veil and the Mask. *Essays on Culture and Ideology. Foreword
  by Ernest Gellner. 140 pp.*
  Zijderfeld, A. C. On Clichés. *The Supersedure of Meaning by Function in Modernity.
  150 pp.*

SOCIOLOGY OF RELIGION

Argyle, Michael and **Beit-Hallahmi, Benjamin.** The Social Psychology of Religion.
  256 pp.
Glasner, Peter E. The Sociology of Secularisation. *A Critique of a Concept.
  146 pp.*
Hall, J. R. The Ways Out. *Utopian Communal Groups in an Age of Babylon. 280 pp.*
Ranson, S., Hinings, B. and **Bryman, A.** Clergy, Ministers and Priests. *216 pp.*
Stark, Werner. The Sociology of Religion. *A Study of Christendom.*
  Volume II. *Sectarian Religion. 368 pp.*
  Volume III. *The Universal Church. 464 pp.*
  Volume IV. *Types of Religious Man. 352 pp.*
  Volume V. *Types of Religious Culture. 464 pp.*
Turner, B. S. Weber and Islam. *216 pp.*
Watt, W. Montgomery. Islam and the Integration of Society. *320 pp.*

SOCIOLOGY OF ART AND LITERATURE

Jarvie, Ian C. Towards a Sociology of the Cinema. *A Comparative Essay on the
  Structure and Functioning of a Major Entertainment Industry. 405 pp.*
Rust, Frances S. Dance in Society. *An Analysis of the Relationships between the Social
  Dance and Society in England from the Middle Ages to the Present Day. 256 pp.
  8 pp. of plates.*
Schücking, L. L. The Sociology of Literary Taste. *112 pp.*
Wolff, Janet. Hermeneutic Philosophy and the Sociology of Art. *150 pp.*

SOCIOLOGY OF KNOWLEDGE

Diesing, P. Patterns of Discovery in the Social Sciences. *262 pp.*

- **Douglas, J. D.** (Ed.) Understanding Everyday Life. *370 pp.*
- **Hamilton, P.** Knowledge and Social Structure. *174 pp.*
  Jarvie, I. C. Concepts and Society. *232 pp.*
  Mannheim, Karl. Essays on the Sociology of Knowledge. *Edited by Paul Kecskemeti. Editorial Note by Adolph Lowe. 353 pp.*
  Remmling, Gunter W. The Sociology of Karl Mannheim. *With a Bibliographical Guide to the Sociology of Knowledge, Ideological Analysis, and Social Planning. 255 pp.*
  Remmling, Gunter W. (Ed.) Towards the Sociology of Knowledge. *Origin and Development of a Sociological Thought Style. 463 pp.*
  Scheler, M. Problems of a Sociology of Knowledge. *Trans. by M. S. Frings. Edited and with an Introduction by K. Stikkers. 232 pp.*

URBAN SOCIOLOGY

Aldridge, M. The British New Towns. *A Programme Without a Policy. 232 pp.*
Ashworth, William. The Genesis of Modern British Town Planning: *A Study in Economic and Social History of the Nineteenth and Twentieth Centuries. 288 pp.*
Brittan, A. The Privatised World. *196 pp.*
Cullingworth, J. B. Housing Needs and Planning Policy: *A Restatement of the Problems of Housing Need and 'Overspill' in England and Wales. 232 pp. 44 tables. 8 maps.*
Dickinson, Robert E. City and Region: *A Geographical Interpretation. 608 pp. 125 figures.*
  The West European City: *A Geographical Interpretation. 600 pp. 129 maps. 29 plates.*
Humphreys, Alexander J. New Dubliners: *Urbanization and the Irish Family. Foreword by George C. Homans. 304 pp.*
Jackson, Brian. Working Class Community: *Some General Notions raised by a Series of Studies in Northern England. 192 pp.*
- **Mann, P. H.** An Approach to Urban Sociology. *240 pp.*
  Mellor, J. R. Urban Sociology in an Urbanized Society. *326 pp.*
  Morris, R. N. and **Mogey, J.** The Sociology of Housing. *Studies at Berinsfield. 232 pp. 4 pp. plates.*
  Mullan, R. Stevenage Ltd. *About 250 pp.*
  Rex, J. and **Tomlinson, S.** Colonial Immigrants in a British City. *A Class Analysis. 368 pp.*
  Rosser, C. and **Harris, C.** The Family and Social Change. *A Study of Family and Kinship in a South Wales Town. 352 pp. 8 maps.*
- **Stacey, Margaret, Batsone, Eric, Bell, Colin** and **Thurcott, Anne.** Power, Persistence and Change. *A Second Study of Banbury. 196 pp.*

RURAL SOCIOLOGY

Mayer, Adrian C. Peasants in the Pacific. *A Study of Fiji Indian Rural Society. 248 pp. 20 plates.*
Williams, W. M. The Sociology of an English Village: *Gosforth. 272 pp. 12 figures. 13 tables.*

SOCIOLOGY OF INDUSTRY AND DISTRIBUTION

Dunkerley, David. The Foreman. *Aspects of Task and Structure. 192 pp.*
Eldridge, J. E. T. Industrial Disputes. *Essays in the Sociology of Industrial Relations. 288 pp.*
Hollowell, Peter G. The Lorry Driver. *272 pp.*
- **Oxaal, I., Barnett, T.** and **Booth, D.** (Eds) Beyond the Sociology of Development.

Economy and Society in Latin America and Africa. 295 pp.

Smelser, Neil J. Social Change in the Industrial Revolution: *An Application of Theory to the Lancashire Cotton Industry, 1770–1840. 468 pp. 12 figures. 14 tables.*

Watson, T. J. The Personnel Managers. *A Study in the Sociology of Work and Employment, 262 pp.*

ANTHROPOLOGY

Brandel-Syrier, Mia. Reeftown Elite. *A Study of Social Mobility in a Modern African Community on the Reef. 376 pp.*

Dickie-Clark, H. F. The Marginal Situation. *A Sociological Study of a Coloured Group. 236 pp.*

Dube, S. C. Indian Village. *Foreword by Morris Edward Opler. 276 pp. 4 plates.*
India's Changing Villages: *Human Factors in Community Development. 260 pp. 8 plates. 1 map.*

Fei, H.-T. Peasant Life in China. *A Field Study of Country Life in the Yangtze Valley. With a foreword by Bronislaw Malinowski. 328 pp. 16 pp. plates.*

Firth, Raymond. Malay Fishermen. *Their Peasant Economy. 420 pp. 17 pp. plates.*

Gulliver, P. H. Social Control in an African Society: a Study of the Arusha, Agricultural Masai of Northern Tanganyika. *320 pp. 8 plates. 10 figures.*
Family Herds. *288 pp.*

Jarvie, Ian C. The Revolution in Anthropology. *268 pp.*

Little, Kenneth L. Mende of Sierra Leone. *308 pp. and folder.*
Negroes in Britain. *With a New Introduction and Contemporary Study by Leonard Bloom. 320 pp.*

Tambs-Lyche, H. London Patidars. *About 180 pp.*

Madan, G. R. Western Sociologists on Indian Society. *Marx, Spencer, Weber, Durkheim, Pareto. 384 pp.*

Mayer, A. C. Peasants in the Pacific. *A Study of Fiji Indian Rural Society. 248 pp.*

Meer, Fatima. Race and Suicide in South Africa. *325 pp.*

Smith, Raymond T. The Negro Family in British Guiana: *Family Structure and Social Status in the Villages. With a Foreword by Meyer Fortes. 314 pp. 8 plates. 1 figure. 4 maps.*

SOCIOLOGY AND PHILOSOPHY

Adriaansens, H. Talcott Parsons and the Conceptual Dilemma. *About 224 pp.*

Barnsley, John H. The Social Reality of Ethics. *A Comparative Analysis of Moral Codes. 448 pp.*

Diesing, Paul. Patterns of Discovery in the Social Sciences. *362 pp.*

● **Douglas, Jack D.** (Ed.) Understanding Everyday Life. *Toward the Reconstruction of Sociological Knowledge. Contributions by Alan F. Blum, Aaron W. Cicourel, Norman K. Denzin, Jack D. Douglas, John Heeren, Peter McHugh, Peter K. Manning, Melvin Power, Matthew Speier, Roy Turner. D. Lawrence Wieder, Thomas P. Wilson and Don H. Zimmerman. 370 pp.*

Gorman, Robert A. The Dual Vision. *Alfred Schutz and the Myth of Phenomenological Social Science. 240 pp.*

Jarvie, Ian C. Concepts and Society. *216 pp.*

Kilminster, R. Praxis and Method. *A Sociological Dialogue with Lukács, Gramsci and the Early Frankfurt School. 334 pp.*

● **Pelz, Werner.** The Scope of Understanding in Sociology. *Towards a More Radical Reorientation in the Social Humanistic Sciences. 283 pp.*

Roche, Maurice. Phenomenology, Language and the Social Sciences. *371 pp.*

Sahay, Arun. Sociological Analysis. *212 pp.*

● **Slater, P.** Origin and Significance of the Frankfurt School. *A Marxist Perspective. 185 pp.*

Spurling, L. Phenomenology and the Social World. *The Philosophy of Merleau-Ponty and its Relation to the Social Sciences. 222 pp.*

Wilson, H. T. The American Ideology. *Science, Technology and Organization as Modes of Rationality. 368 pp.*

International Library of Anthropology
General Editor Adam Kuper

● Ahmed, A. S. Millennium and Charisma Among Pathans. *A Critical Essay in Social Anthropology. 192 pp.*
   Pukhtun Economy and Society. *Traditional Structure and Economic Development. About 360 pp.*

Barth, F. Selected Essays. *Volume I. About 250 pp.* Selected Essays. *Volume II. About 250 pp.*

Brown, Paula. The Chimbu. *A Study of Change in the New Guinea Highlands. 151 pp.*

Foner, N. Jamaica Farewell. *200 pp.*

Gudeman, Stephen. Relationships, Residence and the Individual. *A Rural Panamanian Community. 288 pp. 11 plates, 5 figures, 2 maps, 10 tables.*
   The Demise of a Rural Economy. *From Subsistence to Capitalism in a Latin American Village. 160 pp.*

Hamnett, Ian. Chieftainship and Legitimacy. *An Anthropological Study of Executive Law in Lesotho. 163 pp.*

Hanson, F. Allan. Meaning in Culture. *127 pp.*

Hazan, H. The Limbo People. *A Study of the Constitution of the Time Universe Among the Aged. About 192 pp.*

Humphreys, S. C. Anthropology and the Greeks. *288 pp.*

Karp, I. Fields of Change Among the Iteso of Kenya. *140 pp.*

Lloyd, P. C. Power and Independence. *Urban Africans' Perception of Social Inequality. 264 pp.*

Parry, J. P. Caste and Kinship in Kangra. *352 pp. Illustrated.*

Pettigrew, Joyce. Robber Noblemen. *A Study of the Political System of the Sikh Jats. 284 pp.*

Street, Brian V. The Savage in Literature. *Representations of 'Primitive' Society in English Fiction, 1858–1920. 207 pp.*

Van Den Berghe, Pierre L. Power and Privilege at an African University. *278 pp.*

International Library of Phenomenology and Moral Sciences
General Editor John O'Neill

Apel, K.-O. Towards a Transformation of Philosophy. *308 pp.*

Bologh, R. W. Dialectical Phenomenology. *Marx's Method. 287 pp.*

Fekete, J. The Critical Twilight. *Explorations in the Ideology of Anglo-American Literary Theory from Eliot to McLuhan. 300 pp.*

Medina, A. Reflection, Time and the Novel. *Towards a Communicative Theory of Literature. 143 pp.*

International Library of Social Policy
General Editor Kathleen Jones

Bayley, M. Mental Handicap and Community Care. *426 pp.*

Bottoms, A. E. and McClean, J. D. Defendants in the Criminal Process. *284 pp.*

Bradshaw, J. The Family Fund. *An Initiative in Social Policy. About 224 pp.*

Butler, J. R. Family Doctors and Public Policy. *208 pp.*
Davies, Martin. Prisoners of Society. *Attitudes and Aftercare. 204 pp.*
Gittus, Elizabeth. Flats, Families and the Under-Fives. *285 pp.*
Holman, Robert. Trading in Children. *A Study of Private Fostering. 355 pp.*
Jeffs, A. Young People and the Youth Service. *160 pp.*
Jones, Howard and Cornes, Paul. Open Prisons. *288 pp.*
Jones, Kathleen. History of the Mental Health Service. *428 pp.*
Jones, Kathleen with **Brown, John, Cunningham, W. J., Roberts, Julian** and
   Williams, Peter. Opening the Door. *A Study of New Policies for the Mentally
   Handicapped. 278 pp.*
Karn, Valerie. Retiring to the Seaside. *400 pp. 2 maps. Numerous tables.*
King, R. D. and **Elliot, K. W.** Albany: Birth of a Prison—End of an Era. *394 pp.*
Thomas, J. E. The English Prison Officer since 1850: *A Study in Conflict. 258 pp.*
Walton, R. G. Women in Social Work. *303 pp.*
● **Woodward, J.** To Do the Sick No Harm. *A Study of the British Voluntary Hospital
   System to 1875. 234 pp.*

International Library of Welfare and Philosophy
General Editors Noel Timms and David Watson

● **McDermott, F. E.** (Ed.) Self-Determination in Social Work. *A Collection of Essays
   on Self-determination and Related Concepts by Philosophers and Social Work
   Theorists. Contributors: F. P. Biestek, S. Bernstein, A. Keith-Lucas, D. Sayer,
   H. H. Perelman, C. Whittington, R. F. Stalley, F. E. McDermott, I. Berlin, H. J.
   McCloskey, H. L. A. Hart, J. Wilson, A. I. Melden, S. I. Benn. 254 pp.*
● **Plant, Raymond.** Community and Ideology. *104 pp.*
Ragg, Nicholas M. People Not Cases. *A Philosophical Approach to Social Work.
   168 pp.*
● **Timms, Noel** and **Watson, David.** (Eds) Talking About Welfare. *Readings in
   Philosophy and Social Policy. Contributors: T. H. Marshall, R. B. Brandt, G. H.
   von Wright, K. Nielsen, M. Cranston, R. M. Titmuss, R. S. Downie, E. Telfer, D.
   Donnison, J. Benson, P. Leonard, A. Keith-Lucas, D. Walsh, I. T. Ramsey.
   320 pp.*
● Philosophy in Social Work. *250 pp.*
● **Weale, A.** Equality and Social Policy. *164 pp.*

Library of Social Work
General Editor Noel Timms

● **Baldock, Peter.** Community Work and Social Work. *140 pp.*
○ **Beedell, Christopher.** Residential Life with Children. *210 pp. Crown 8vo.*
● **Berry, Juliet.** Daily Experience in Residential Life. *A Study of Children and their
   Care-givers. 202 pp.*
○ Social Work with Children. *190 pp. Crown 8vo.*
● **Brearley, C. Paul.** Residential Work with the Elderly. *116 pp.*
● Social Work, Ageing and Society. *126 pp.*
● **Cheetham, Juliet.** Social Work with Immigrants. *240 pp. Crown 8vo.*
● **Cross, Crispin P.** (Ed.) Interviewing and Communication in Social Work.
   *Contributions by C. P. Cross, D. Laurenson, B. Strutt, S. Raven. 192 pp. Crown
   8vo.*

● **Curnock, Kathleen** and **Hardiker, Pauline.** Towards Practice Theory. *Skills and Methods in Social Assessments. 208 pp.*

● **Davies, Bernard.** The Use of Groups in Social Work Practice. *158 pp.*

● **Davies, Martin.** Support Systems in Social Work. *144 pp.*

Ellis, June. (Ed.) West African Families in Britain. *A Meeting of Two Cultures. Contributions by Pat Stapleton, Vivien Biggs. 150 pp. 1 Map.*

● **Hart, John.** Social Work and Sexual Conduct. *230 pp.*

● **Hutten, Joan M.** Short-Term Contracts in Social Work. *Contributions by Stella M. Hall, Elsie Osborne, Mannie Sher, Eva Sternberg, Elizabeth Tuters. 134 pp.*

Jackson, Michael P. and **Valencia, B. Michael.** Financial Aid Through Social Work. *140 pp.*

● **Jones, Howard.** The Residential Community. *A Setting for Social Work. 150 pp.*

● (Ed.) Towards a New Social Work. *Contributions by Howard Jones, D. A. Fowler, J. R. Cypher, R. G. Walton, Geoffrey Mungham, Philip Priestley, Ian Shaw, M. Bartley, R. Deacon, Irwin Epstein, Geoffrey Pearson. 184 pp.*

Jones, Ray and **Pritchard, Colin.** (Eds) Social Work With Adolescents. *Contributions by Ray Jones, Colin Pritchard, Jack Dunham, Florence Rossetti, Andrew Kerslake, John Burns, William Gregory, Graham Templeman, Kenneth E. Reid, Audrey Taylor. About 170 pp.*

○ **Jordon, William.** The Social Worker in Family Situations. *160 pp. Crown 8vo.*

● **Laycock, A. L.** Adolescents and Social Work. *128 pp. Crown 8vo.*

● **Lees, Ray.** Politics and Social Work. *128 pp. Crown 8vo.*

● Research Strategies for Social Welfare. *112 pp. Tables.*

○ **McCullough, M. K.** and **Ely, Peter J.** Social Work with Groups. *127 pp. Crown 8vo.*

● **Moffett, Jonathan.** Concepts in Casework Treatment. *128 pp. Crown 8vo.*

Parsloe, Phyllida. Juvenile Justice in Britain and the United States. *The Balance of Needs and Rights. 336 pp.*

● **Plant, Raymond.** Social and Moral Theory in Casework. *112 pp. Crown 8vo.*

Priestley, Philip, Fears, Denise and **Fuller, Roger.** Justice for Juveniles. *The 1969 Children and Young Persons Act: A Case for Reform? 128 pp.*

● **Pritchard, Colin** and **Taylor, Richard.** Social Work: Reform or Revolution? *170 pp.*

○ **Pugh, Elisabeth.** Social Work in Child Care. *128 pp. Crown 8vo.*

● **Robinson, Margaret.** Schools and Social Work. *282 pp.*

○ **Ruddock, Ralph.** Roles and Relationships. *128 pp. Crown 8vo.*

● **Sainsbury, Eric.** Social Diagnosis in Casework. *118 pp. Crown 8vo.*

● Social Work with Families. *Perceptions of Social Casework among Clients of a Family Service. 188 pp.*

Seed, Philip. The Expansion of Social Work in Britain. *128 pp. Crown 8vo.*

● **Shaw, John.** The Self in Social Work. *124 pp.*

Smale, Gerald G. Prophecy, Behaviour and Change. *An Examination of Self-fulfilling Prophecies in Helping Relationships. 116 pp. Crown 8vo.*

Smith, Gilbert. Social Need. *Policy, Practice and Research. 155 pp.*

● Social Work and the Sociology of Organisations. *124 pp. Revised edition.*

● **Sutton, Carole.** Psychology for Social Workers and Counsellors. *An Introduction. 248 pp.*

● **Timms, Noel.** Language of Social Casework. *122 pp. Crown 8vo.*

● Recording in Social Work. *124 pp. Crown 8vo.*

● **Todd, F. Joan.** Social Work with the Mentally Subnormal. *96 pp. Crown 8vo.*

● **Walrond-Skinner, Sue.** Family Therapy. *The Treatment of Natural Systems. 172 pp.*

● **Warham, Joyce.** An Introduction to Administration for Social Workers. *Revised edition. 112 pp.*

● An Open Case. *The Organisational Context of Social Work. 172 pp.*

○ **Wittenberg, Isca Salzberger.** Psycho-Analytic Insight and Relationships. *A Kleinian Approach. 196 pp. Crown 8vo.*

Primary Socialization, Language and Education
General Editor   Basil Bernstein

Adlam, Diana S., *with the assistance of Geoffrey Turner and Lesley Lineker.* Code in *Context. 272 pp.*
Bernstein, Basil. Class, Codes and Control. *3 volumes.*
● 1. *Theoretical Studies Towards a Sociology of Language. 254 pp.*
2. *Applied Studies Towards a Sociology of Language. 377 pp.*
● 3. *Towards a Theory of Educational Transmission. 167 pp.*
Brandis, W. and **Bernstein, B.** Selection and Control. *176 pp.*
Brandis, Walter and **Henderson, Dorothy.** Social Class, Language and Communication. *288 pp.*
Cook-Gumperz, Jenny. Social Control and Socialization. *A Study of Class Differences in the Language of Maternal Control. 290 pp.*
● **Gahagan, D. M.** and **G. A.** Talk Reform. *Exploration in Language for Infant School Children. 160 pp.*
Hawkins, P. R. Social Class, the Nominal Group and Verbal Strategies. *About 220 pp.*
Robinson, W. P. and **Rackstraw, Susan D. A.** A Question of Answers. *2 volumes. 192 pp. and 180 pp.*
Turner, Geoffrey J. and **Mohan, Bernard A.** A Linguistic Description and Computer Programme for Children's Speech. *208 pp.*

Reports of the Institute of Community Studies

Baker, J. The Neighbourhood Advice Centre. A Community Project in Camden. *320 pp.*
● **Cartwright, Ann.** Patients and their Doctors. *A Study of General Practice. 304 pp.*
Dench, Geoff. Maltese in London. *A Case-study in the Erosion of Ethnic Consciousness. 302 pp.*
Jackson, Brian and **Marsden, Dennis.** Education and the Working Class: *Some General Themes Raised by a Study of 88 Working-class Children in a Northern Industrial City. 268 pp. 2 folders.*
Marris, Peter. The Experience of Higher Education. *232 pp. 27 tables.*
● Loss and Change. *192 pp.*
Marris, Peter and **Rein, Martin.** Dilemmas of Social Reform. *Poverty and Community Action in the United States. 256 pp.*
Marris, Peter and **Somerset, Anthony.** African Businessmen. *A Study of Entrepreneurship and Development in Kenya. 256 pp.*
Mills, Richard. Young Outsiders: *a Study in Alternative Communities. 216 pp.*
Runciman, W. G. Relative Deprivation and Social Justice. *A Study of Attitudes to Social Inequality in Twentieth-Century England. 352 pp.*
Willmott, Peter. Adolescent Boys in East London. *230 pp.*
Willmott, Peter and **Young, Michael.** Family and Class in a London Suburb. *202 pp. 47 tables.*
Young, Michael and **McGeeney, Patrick.** Learning Begins at Home. *A Study of a Junior School and its Parents. 128 pp.*
Young, Michael and **Willmott, Peter.** Family and Kinship in East London. *Foreword by Richard M. Titmuss. 252 pp. 39 tables.*
The Symmetrical Family. *410 pp.*

Reports of the Institute for Social Studies in Medical Care

Cartwright, Ann, Hockey, Lisbeth and **Anderson, John J.** Life Before Death. *310 pp.*
Dunnell, Karen and **Cartwright, Ann.** Medicine Takers, Prescribers and Hoarders. *190 pp.*
Farrell, C. My Mother Said. . . *A Study of the Way Young People Learned About Sex and Birth Control. 288 pp.*

Medicine, Illness and Society
General Editor W. M. Williams

Hall, David J. Social Relations & Innovation. *Changing the State of Play in Hospitals. 232 pp.*
Hall, David J. and **Stacey, M.** (Eds) Beyond Separation. *234 pp.*
Robinson, David. The Process of Becoming Ill. *142 pp.*
Stacey, Margaret *et al.* Hospitals, Children and Their Families. *The Report of a Pilot Study. 202 pp.*
Stimson, G. V. and **Webb, B.** Going to See the Doctor. *The Consultation Process in General Practice. 155 pp.*

Monographs in Social Theory
General Editor Arthur Brittan

● **Barnes, B.** Scientific Knowledge and Sociological Theory. *192 pp.*
Bauman, Zygmunt. Culture as Praxis. *204 pp.*
● **Dixon, Keith.** Sociological Theory. *Pretence and Possibility. 142 pp.*
The Sociology of Belief. *Fallacy and Foundation. About 160 pp.*
Goff, T. W. Marx and Mead. *Contributions to a Sociology of Knowledge. 176 pp.*
Meltzer, B. N., Petras, J. W. and **Reynolds, L. T.** Symbolic Interactionism. *Genesis, Varieties and Criticisms. 144 pp.*
● **Smith, Anthony D.** The Concept of Social Change. *A Critique of the Functionalist Theory of Social Change. 208 pp.*

Routledge Social Science Journals

The British Journal of Sociology. *Editor – Angus Stewart; Associate Editor – Leslie Sklair. Vol. 1, No. 1 – March 1950 and Quarterly. Roy. 8vo. All back issues available. An international journal publishing original papers in the field of sociology and related areas.*
Community Work. *Edited by David Jones and Marjorie Mayo. 1973. Published annually.*
Economy and Society. *Vol. 1, No. 1. February 1972 and Quarterly. Metric Roy. 8vo. A journal for all social scientists covering sociology, philosophy, anthropology, economics and history. All back numbers available.*

Ethnic and Racial Studies. *Editor – John Stone. Vol. 1 – 1978. Published quarterly.*
Religion. Journal of Religion and Religions. *Chairman of Editorial Board, Ninian Smart. Vol. 1, No. 1, Spring 1971. A journal with an inter-disciplinary approach to the study of the phenomena of religion. All back numbers available.*
Sociology of Health and Illness. *A Journal of Medical Sociology. Editor – Alan Davies; Associate Editor – Ray Jobling. Vol. 1, Spring 1979. Published 3 times per annum.*
Year Book of Social Policy in Britain. *Edited by Kathleen Jones. 1971. Published annually.*

Social and Psychological Aspects of Medical Practice
Editor Trevor Silverstone

Lader, Malcolm. Psychophysiology of Mental Illness. *280 pp.*
● **Silverstone, Trevor** and **Turner, Paul.** Drug Treatment in Psychiatry. *Revised edition. 256 pp.*
Whiteley, J. S. and **Gordon, J.** Group Approaches in Psychiatry. *240 pp.*